# We Came In Peace For All Mankind

# We Came In Peace For All Mankind

## The Untold Story of the Apollo 11 Silicon Disc

**By Tahir Rahman**

Produced by Tahir Rahman
Designed by Michele Rook
Cover Designed by Angela Farley
All photographs courtesy of NASA and John Sprague

Silicon Disc images © Tahir Rahman

All rights reserved. No part of this publication may be reproduced or transmitted in any form or by any means, electronic or mechanical, including photocopying, recording or any information storage and retrieval system, without permission from copyright holder.

Copyright © 2008 by Tahir Rahman
First Edition — First Printing
Printed in the USA

ISBN 978-1-58597-441-2
Library of Congress Control Number: 2007937314

4500 College Boulevard
Overland Park, Kansas 66211
888-888-7696
www.leatherspublishing.com

**The silicon disc represents an historic time when many nations looked beyond their differences to come together to achieve this historic first.**

— Charlie Duke, Apollo 16 Moonwalker

*We Came in Peace* and we will return with hope and peace for all mankind.

— Gene Cernan, Apollo 17 Moonwalker

**A Great Idea!**

— Scott Carpenter, Mercury astronaut

*With love to my wife, Stephanie  
and our children, Jacqueline and Alec.*

*In the spirit of knowledge, peace & hope for all mankind.*

*Tariq Rahman*
*2/12/2009*

*It is our earnest hope for mankind that while we gain the Moon, we shall not lose the world.*

**— Eric Williams, Prime Minister of Trinidad and Tobago**
*one of 74 goodwill messages on the tiny Apollo 11 silicon disc.*

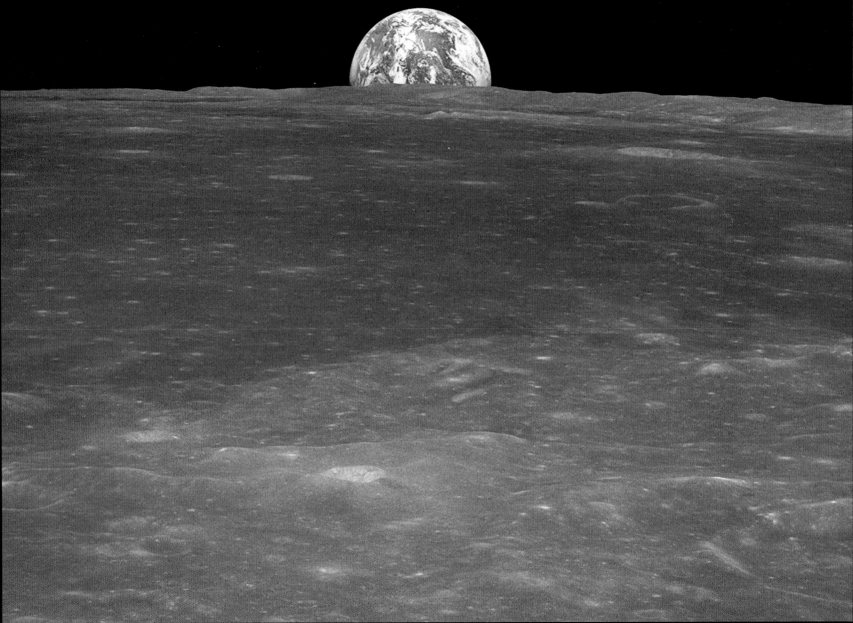

Neil Armstrong peered through one of the small windows of the lunar module, Eagle. He and Buzz Aldrin saw something that no human beings had seen before- the forbidding lunar surface. Armstrong described the scene to Houston: "It's pretty much without color. It's gray, and it's very white, chalky gray as you look into the zero-phase line. And it's considerably darker gray, more like ashen gray, as you look out ninety degrees to the Sun." The astronauts had just descended 50,000 feet from lunar orbit and avoided flying into a giant crater. Now, instead of taking a scheduled nap, Armstrong and Aldrin would climb out of the lunar module and complete their historic mission.

An estimated 600 million people watched the first lunar landing.

People cheered, cried, laughed, and lit up cigars. The world united in a unique way while the astronauts walked on a surreal world for the first time in human history. Here are the untold stories of how a special U.S. flag, a lunar plaque, and a tiny silicon disc with beautiful and powerful words traveled with a trio of astronauts to the Sea of Tranquility.

*A rare photo of Neil Armstrong on the Moon (See full panoramic on pages 70-71).*

## A forward step for all mankind

Those planning the first lunar landing knew that once the astronauts landed on the Moon, they would spend most of their time doing scientific research, gathering samples, and taking photographs, but the landing would be viewed worldwide. The United States would be the first country to set foot on the Moon, and the astronauts would represent the entire planet. It was determined, just months prior to launch, that some kind of ceremony should take place on the Moon.

In February 1969, Willis H. Shapley, NASA associate deputy administrator, was appointed chairman of a committee to plan symbolic activities for the astronauts to perform on the first lunar landing. Their conclusions and the resulting plans included:

1. No activity should jeopardize crew safety.
2. The activities should be in good taste from a world perspective.
3. An historic "forward step for all mankind" theme.
4. Make it clear that this was an American accomplishment, symbolized by placing the U.S. flag on the surface, without implying U.S. sovereignty on the Moon.
5. A commemorative plaque, affixed to the lunar module descent stage, would include the two hemispheres of the Earth, without boundaries. It would also contain the names of the astronauts and the president of the United States. There would be a short and simple statement regarding the peaceful nature of the mission, and that the achievement was for all mankind.
6. Miniature flags of all U.S. states, the District of Columbia, U.S. territories, flags of all nations (to be presented to heads of state), and miniature U.S. flags, as well as two full size U.S. flags (for the House and Senate), were to be carried in the command module only.
7. Create a stamp die from which the U.S. post office would print commemorative stamps.

By June 8, with just over a month prior to launch, U. Alexis Johnson, Deputy Undersecretary for Political Affairs at the U.S. State Department, recommended that a plaque be placed on the Moon. He drafted a statement that read:

> We who first walk the surface of the moon leave this plaque to commemorate our journey and to mark Man's progress in his continuing quest for a more complete understanding of the universe. We came as envoys of mankind, exploring the moon for the benefit of all peoples. May this voyage not only illuminate the mysteries of the universe, but unite us in the search for truth and understanding on our own planet.
>
> Names of Astronauts
> Date

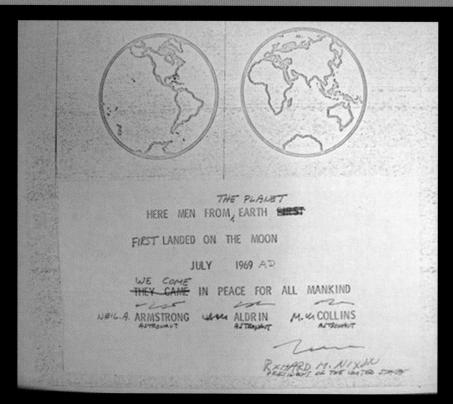

*Early version of plaque with handwritten changes made.*

**The plaque was later changed to read:**

HERE MEN FROM EARTH

LANDED ON THE MOON

JULY, 1969 A.D.

THEY CAME IN PEACE FOR ALL MANKIND

ARMSTRONG    ALDRIN    COLLINS

*An early version of the plaque.*

HERE MEN FROM THE PLANET EARTH
FIRST SET FOOT UPON THE MOON
JULY 1969, A.D.
WE CAME IN PEACE FOR ALL MANKIND

NEIL A. ARMSTRONG
ASTRONAUT

MICHAEL COLLINS
ASTRONAUT

EDWIN E. ALDRIN, JR.
ASTRONAUT

RICHARD NIXON
PRESIDENT, UNITED STATES OF AMERICA

HERITAGE AMERICANA

*Final version of plaque.*

Julian Sheer, head of the Public affairs for NASA fought over the wording of the plaque. One of President Nixon's staff insisted that the words "under God" be added. Sheer argued that this might offend other religions. Sheer agreed to add these words, despite knowing that it was too late to change it. He later wrote, "It occurred to me that in the rush of events, no one would remember."

The plaque was made of stainless steel measuring 9 x 7⅝ inches and ¹⁄₁₆ inches thick. The world hemispheres and words were filled in with black epoxy. It was bent around a four inch radius to fit around a strut. A thin sheet of steel protected the plaque and would be removed by Neil Armstrong.

# The Apollo 11 Crew

*From left to right: Neil Armstrong, Michael Collins and Buzz Aldrin.*

## From Kitty Hawk to the Moon

On December 14, 1903, Orville and Wilbur Wright flipped a coin to see who would make the first attempt at powered flight. On December 17, Orville made an historic first powered flight. His first flight lasted 12 seconds for a total distance of 120 feet. The flyer was slightly damaged that same day. Fragments of the wood and fabric from the Wright flyer would be carried to the Moon and brought back for display in a museum.

## Lunar flag issues

There was controversy surrounding the planting of a flag on the Moon. Many were concerned about the perception that the U.S. was acquiring the Moon for itself. Some suggested that a United Nations (U.N.) flag be planted by the astronauts. Alexis Johnson indicated that the use of the United Nations flag would not be adequate, because it did not represent all the countries of the world. The U.N. did not yet represent West Germany, South Vietnam, or South Korea. He went on to state that "The raising of an American flag would seem most undesirable from this standpoint, since such an action has historically symbolized conquest and territorial acquisition."

At this point, the U.S. Congress intervened and decided this was an American project and that a U.S. flag would be planted on the Moon.

> *My job was to get the flag there. I was less concerned about whether that was the right artifact to place. I let other, wiser minds than mine make those kinds of decisions, and I had no problem with it.*
>
> —Neil Armstrong

Jack Kinzler, Chief of Technical Services at the Manned Spacecraft Center (MSC) was given the honor of designing America's flag to fly to the Moon. Kinzler decided that a full-sized flag could be designed and subsequently planted on the Moon by the astronauts. Several engineering considerations were taken into account:

1. The lunar module had limited external storage space. The flag had to be accessible to the astronauts once they were on lunar soil, therefore attaching it to the outside of the module made sense.
2. The lunar module engine would experience temperature extremes during the descent phase of the landing. This could harm the flag.
3. A limited weight would be necessary.
4. Unpacking and assembling the flag should be simple and within the constraints of the astronauts' limited range of motion in their bulky space suits.
5. The flag should appear to be flowing in a breeze. The Moon would have no wind and very light gravity.

The engineers on Kinzler's team responded with some creative solutions to the problems. They designed a 3 x 5 foot version of the Stars and Stripes that weighed 9 pounds, 7 ounces. It was made of nylon, and had a telescoping arm that would attach to the top edge where a hem was sewn.

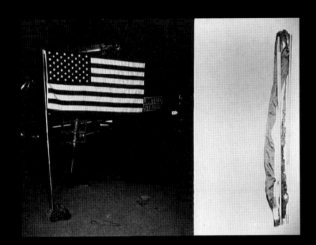 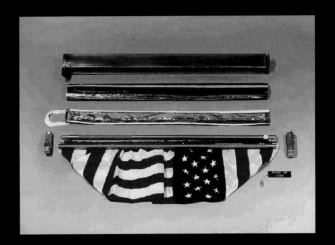

   The flag would be mounted on the left side of the lunar module, near the ladder. This would allow the astronauts to access it easily from outside the spacecraft. The flag would be unfurled by extending the telescoping arm. A catch would prevent the hinge between the telescoping arm and the pole from moving. A base portion of the flagpole with a pointed tip would be driven into the ground. The main flagpole would then be slipped into the base pole. A red ring of paint, eighteen inches from the bottom of the flagpole, was used as a marking point to estimate how far into the lunar soil the pole had penetrated.
   For temperature extremes, a shroud protected the flag assembly. Thermoflex insulation was used between the aluminum and a stainless steel outer case. Thermal blanketing material was placed in layers between the shroud and flag assembly material. Velcro held the flag packaging together. A pull tab was installed to allow easy opening of the assembly by the astronauts.

# The flag is rushed to the Saturn V rocket

The ceremonial activities were decided with very little time remaining before launch. As Kinzler prepared his special lunar flag, the powerful Saturn V rocket sat at launch pad 39-A. By now, the space center had been named after John F. Kennedy, the nation's martyred president, who had given NASA a nearly impossible timetable to complete the Moon landing. With very little time left before the launch, the plaque and flag assembly were placed on the spidery looking spacecraft called the lunar module.

*The ladder of the Lunar Module is seen here with the covered plaque. Armstrong would unveil the plaque after climbing down this ladder.*

## Presidential Moon dust orders

President Richard Nixon requested that small samples of lunar dust be delivered to nations of the world. A White House memorandum dated June 2, 1969, for NASA administrator Thomas Paine, stated that Nixon was "captivated by the idea of even a miniscule amount of Moon dust, appropriately contained and mounted, as a present for all of the foreign heads of state." Political power had already influenced the distribution of some tiny lunar samples. Scientists still had no idea what the Moon was made of, and whether or not it was safe.

## The rocket's red glare

President Nixon also wanted the national anthem played while the astronauts were on the lunar surface, but that idea was vetoed by NASA's special presidential advisor for Apollo 11, Frank Borman. A former astronaut, Colonel Borman was the commander of Apollo 8, the mission that put humans into lunar orbit for the first time.

Borman argued that the symbolic activities were already taking up so much time in the schedule that additional time would be taken away from scientific discoveries. He also pointed out that "continuous modulation of carrier waves for two and half minutes could prove to be a potential hazard for the crew's well-being." Borman went on to recommend that the national anthem be played upon the crew's return to Earth.

## World leaders are contacted by NASA

In June, 1969, NASA originated the concept of gathering goodwill messages from foreign heads of state. This was a last minute decision. It was also a somewhat risky public relations issue. Not every country liked America or its foreign policy. NASA Administrator Tom Paine and Assistant Secretary Alexis Johnson of the U.S. State Department discussed the idea by telephone on June 27. In an unprecedented event, NASA was authorized to contact foreign delegations with a request for chiefs of state to provide a message for deposit on the Moon.

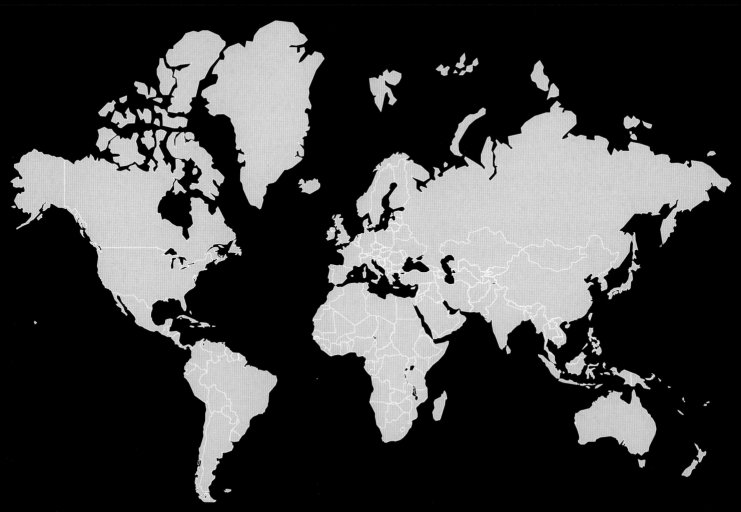

## A letter to nations of the world read:

*Dear Mr. Ambassador:*

*NASA has received from several nations messages of good will for the Apollo 11 astronauts to deposit on the Moon. If your Chief of State would also like to have his message taken to the Moon by the Apollo astronauts, we would be pleased to make the necessary arrangements. To meet our countdown schedule, it will be necessary to receive a document suitable for microfilming. It is intended to deposit the microfilmed documents on the Moon in conjunction with a plaque signed by the President of the United States and the three astronauts, which states, "Here men from the Planet Earth first set foot upon the Moon. We came in peace for all mankind."*

NASA sent 116 requests. They received 81 replies, of which 73 provided messages from heads of state. The message from Venezuela arrived too late. Seven nations expressed an inability to provide messages for various reasons. Germany declined due to a change in their government. Many communist countries, including the Soviet Union and eastern block of Europe, declined or did not respond. The U.S. was at war in Vietnam. The Middle East was embroiled in regional conflicts. A cold war existed between the U.S. and U.S.S.R.

> THE GERMAN AMBASSADOR
> WASHINGTON, D.C.
> July 1, 1969
>
> Dear Dr. Paine:
>
> This is in reference to your kind letter of June 23, 1969, in which you indicated the opportunity to have a good will message by the German President deposited on the moon together with similar messages of other Chiefs of State.
>
> I have been in constant communication with my Government in order to meet the deadline of June 30 as mentioned in your letter. Due to the change of the Presidency in Germany, however, which took place on July first, it was unfortunately not possible to make the administrative and technical arrangements necessary for transmitting the message to Washington in time.
>
> I regret, therefore, not to be in a position to follow your suggestion which deeply impressed me.
>
> Sincerely yours,
>
> *Rolf Pauls*
>
> Mr. T. O. Paine,
> Administrator
> National Aeronautics and
> Space Administration
> Washington, D.C. 20546

*A response from the German Ambassador. Germany's government, along with several others declined to send a message.*

The nations that were invited to contribute but were *not* represented on the disc included:

*Austria, Barbados, Bolivia, Botswana, Bulgaria, Burma, Burundi, Cameroon, Central African Republic, Ceylon, Czechoslovakia, El Salvador, Finland, France, Gabon, Germany, Guatemala, Guinea, Haiti, Honduras, Hungary, Indonesia, Libya, Lithuania, Luxembourg, Malawi, Niger, Nigeria, Paraguay, Poland, Rwanda, Saudi Arabia, Singapore, Somalia, Spain, Sweden, Switzerland, Tanzania, Uganda, USSR, and Venezuela.*

Since there was such little time to respond to the messages, some leaders expressed confusion over the matter. For example, the King of Thailand sent a telegram on launch day (July 16, 1969) to the U.S. State Department stating:

"In view of our total ignorance of this project... and King's apparent keen interest, would appreciate any information you can provide concerning NASA invitation to send message . . . number of countries responding . . . methods of recording and method of deposit on the Moon."

# From atomic bombs to goodwill messages — the Sprague Electric Company (1941-1986)

*Sprague Electric Building — Courtesy of John Sprague*

Little did the King of Thailand realize that even NASA didn't know exactly how the messages would get there.

On June 23, 1969, NASA approached the Sprague Electric Company of North Adams, Massachusetts with a crash project. Sprague needed to devise a plan for leaving the goodwill messages from world leaders on the Moon. It needed to be small and lightweight since weight and space were limited in the tiny spacecraft. Ordinary microfilm was out of the question since it would not endure the harsh conditions on the lunar surface: extreme temperatures, radiation, cosmic rays, and solar flares. This "time capsule" of messages needed to last for thousands of years.

This posed an unusual engineering problem, and Sprague was the perfect firm to help. It already had government contracts and years of experience in space-age technology. In the early 1940's, Sprague Electric manufactured capacitors and other types of electronic components, primarily for consumer application like radios and televisions. It also began a clandestine program to make a sophisticated firing capacitor used in the triggering mechanism for the atomic bomb. The company's code name was *Manhattan Square*, a reference to the government's *Manhattan Project*.

Undersecretary of War, Robert P. Patterson, in a message received August 20th, 1945, congratulated Sprague Electric employees who worked on parts of the atomic bomb:

*Today the whole world knows the secret you have helped us keep for many months. I am pleased to be able to add that the warlords of Japan now know its effects better than we ourselves. The atomic bomb, which you have helped to develop with high devotion to patriotic duty, is the most devastating military weapon that any country has ever been able to turn against its enemy. No one of you has worked on the entire project or known the whole story. Each of you has done his own job and kept his own secret, so today I speak for a grateful nation when I say congratulations, and thank you all. I hope you will continue to keep the secrets you have kept so well. The need for security and for continued effort is as fully great now as it ever was. We are proud of every one of you.*

# Silicon: an ideal message palette

As Sprague's involvement in government programs increased, it built untold numbers of components for NASA. There were 53,000 Sprague components on the Apollo 11 spacecraft. Many more were used in ground control equipment. Sprague manufactured components for the Gemini, Apollo, and Skylab missions.

Hard at work to enshrine the goodwill messages, the ingenious engineers of Sprague Electric realized that silicon was the perfect storage medium.

1. Silicon could store large amounts of data in a very small area.
2. Silicon was inherently stable, both to environment and temperature.

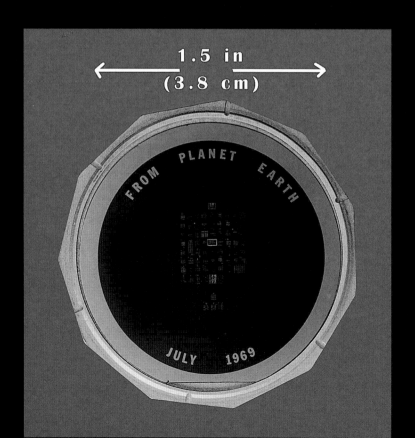

Furthermore, they could leverage existing photolithographic techniques used to etch tiny high-speed integrated circuits for computers and other electronic applications, and complete their task in time for launch.

Following the invention of the transistor in 1948, at the Bell Telephone Laboratories, this technology had evolved into the so-called planar process, which had been introduced by Jean Hoerni of Fairchild Semiconductor in 1958.

Variations of this process are the basis of fabrication for today's integrated circuits. This breakthrough technology would become like an artist's palette for Sprague to manufacture the Moon disc.

> *Crash program is an understatement. We had almost no time to put this together!*
>
> *— John L. Sprague, son of founder Robert C. Sprague and head of Sprague's semiconductor division.*

## U.S. patent number 3,607,347

NASA gave Sprague just weeks to make the disc. Invented by John L. Sprague, Robert S. Pepper, Eugene P. Donovan, and Frederick W. Howe the disc was 3.8 cm (1.5 inches) and appeared to be a dark gray, glassy, and mirror-like material. It was composed of 99.9999% pure silicon. The messages could be readily viewed through any low- powered microscope with just 60-100x magnification.

A person might see the presence of the messages with the naked eye, but would not be able to read the individual words. The disc prominently stated: "From Planet Earth, July, 1969." The tiny messages appeared gold in color and when combined with native language writing, became intricate artwork and poetry.

To make the disc, Sprague engineers photographed full-page paper messages and then reduced them in size 200 times, until the lower case letters were just one-fourth the thickness of a human hair. The shrunken messages were then transferred to a glass stencil, and applied to the disc with a secret emulsion and buffering process, known only to a few (see appendix3).

*It was a rush to get it done. We slept on lab benches for two days in a row.*

*— Ray Carswell, Sprague Engineer*

The messages started arriving at Sprague from NASA during the Independence Day holidays on July 3 and 4. The finished discs were delivered to NASA on July 6. On July 9, an unexpected announcement came from NASA. Additional messages from more countries were sent to NASA and the job had to be completely redone. John Sprague recalled that, "the last minute requirement of a complete redo to add more world messages came as an unexpected shock."

The second and final disc version was delivered to NASA on July 11, just nine days before being placed on the Moon.

The exact number of each version of discs made by Sprague is unknown. A press release stated that seventeen discs were made. They were given to Sprague workers, top leaders at NASA, astronauts, government officers, and Presidents Nixon and Johnson. But only one disc would go to the Moon- and it was placed in an aluminum case, similar to a woman's make-up case, for protection.

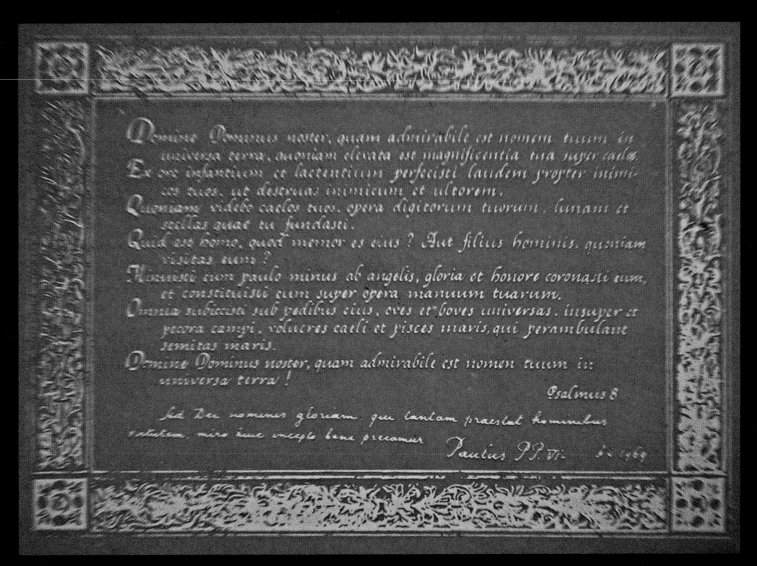

*Message from Pope Paul VI etched on silicon disc (100x magnification).*

John Sprague recalls that a few factors lead to their success. The semiconductor division was still in its adolescence, and being in a small company was an asset. According to Sprague, "We were unencumbered by all the bureaucratic layers of a larger organization. Decisions could be made and implemented instantaneously, and no one even considered failure."

*Robert Sprague seen holding the silicon disc.*
*The photograph he is holding shows the aluminum case that secured the disc.*

# The package

In addition to the various pieces of equipment needed for the lunar excursion, the astronauts took aboard a small package that contained the silicon disc, a gold olive branch, and a few memorial items to honor earlier space explorers. These items would be left on the Moon in a special pouch.

Among the items was something from Apollo 1. The Apollo 1 crew, Gus Grissom, Roger Chaffee, and Ed White, had perished in a horrific fire during a test of their spacecraft in 1967. They asphyxiated while their space suits melted and fused together. Buzz Aldrin recalls with great sadness in his voice, "Ed was a good friend of mine." The fire incident nearly ended the lunar landing program. A congressional investigation and restructuring of NASA took place, and the changes made after the fire paved the way to the successful 1969 lunar landing program.

The fire made everyone more cautious, and certainly invigorated NASA to a higher level of standards and safety. One such development, Beta cloth was invented by the Dow-Corning company to create the astronaut's space suits. This material, made of silicon, does not burn. It will only melt at temperatures exceeding 650 degrees Celsius.

Apollo 1 made a nation mourn the loss of its heroes. Now the ashen dust of the Sea of Tranquility would honor the three men's sacrifice. Inside a Beta cloth pouch, a beautiful Apollo 1 patch would be left on the Moon. The fire resistant pouch's very existence demonstrated that the astronauts did not die in vain.

A gold olive branch, symbolizing peace, was also carried to the Moon. Three other gold olive branches were carried by the astronauts to give to their wives upon their return to Earth. Buzz Aldrin recalls having two gem stones placed on the olive branch that he gave to his wife.

Two commemorative Soviet medals were included in the package. They honored cosmonauts Yuri Gagarin and Vladmir Komarov. Gagarin was the first human being to orbit the Earth. He was later killed in an airplane crash. Komarov was killed when his parachutes failed to open during his Soyuz flight.

## The eagle's talons

Each NASA mission receives a specially designed patch to be worn on uniforms of those involved in the project. Astronaut James Lovell (Apollo 8 and 13) suggested that an eagle be used for the Apollo 11 insignia. Michael Collins, command module pilot, created this beautiful Apollo 11 patch. Collins originally traced an eagle from a *National Geographic* magazine, and drew an olive branch in the eagle's beak, but the government rejected it, fearing it looked too aggressive. The government did not want to render an image of America "taking over" the Moon. International law had already stated that the Moon did not belong to any one country. Instead, the olive branch was placed in the eagle's talons, so it appeared that the eagle was bringing it to the Moon. The word Eagle was also used as the lunar module's (LEM) name.

# We have liftoff!

On July 16, 1969, the massive Saturn V rocket, 363 feet tall, carried Neil Alden Armstrong, Edwin "Buzz" Aldrin, and Michael Collins toward Earth orbit. Armstrong was the commander of this flight, Apollo 11, the mission to put a man on the Moon. The rocket climbed slowly at first, like a majestic candle, and then gained speed. The crew felt the intense vibration of over seven million pounds of thrust. The astronauts experienced a thrilling ride as the controlled explosion hurled them into the heavens. But neatly tucked away in the lunar module storage bay, the tiny silicon disc was safe and sound, awaiting its own role in history just four days later.

Over 300,000 men and women helped build, test, and coordinate this most magnificent machine. Man's greatest adventure was well on the way.

During the Islamic golden age, navigational sciences were developed, making use of a kamal (sextant) to determine the altitudes of stars. Hundreds of years ago, the first sextant was constructed in Iran, by Abu-Mahmud al-Khujandi. Soon detailed maps of the period, along with instruments, allowed sailors to boldly traverse oceans rather than cautiously following the coasts.

The Apollo 11 astronauts used similar instruments to guide themselves through cislunar space. Michael Collins used star charts and made sightings through eyepieces. There were code numbers for the stars, like Rigel in the constellation Orion, the same stars used by the ancient Persian navigators.

The astronauts voyaged further from the Earth and noticed that it shrank in size, giving them a sense of humility that only very few humans have experienced. The photos of that tiny globe are amongst the most celebrated in history. The blackness of empty space surrounding the astronauts evoked strong emotions, something a mere photograph will never do.

> *On this occasion when Neil Armstrong and Colonel Edwin Aldrin set foot for the first time on the Moon from the Earth, we pray the Almighty God to guide mankind towards ever increasing success in the establishment of peace and the progress of culture, knowledge and human civilization.*
>
> — *Mohammad Reza Pahlavi, message from the Shah of Iran etched onto silicon disc*

As they entered lunar orbit in the evening of July 20, Neil Armstrong and Buzz Aldrin climbed into the lunar module *(Eagle)* and departed from the command module *(Columbia)*, leaving the spacecraft in the capable hands of Michael Collins. Collins, alone in his spacecraft, flew over the dark side of the Moon with all of humanity on one side and open space to his other side. He chose the name *Columbia* for the ship he would largely control. *Columbia* was similar to Jules Verne's *Columbiad* cannon that hurled men to the Moon. The name had national symbolic roots, and was derived from the name of Christopher Colombus.

Neil Armstrong and Buzz Aldrin had simulated flying the lunar module many times, but now it was for real. This time they were a quarter of a million miles from Earth. They stood in *Eagle* which was designed to be flown without seats. This was done to keep its weight to a minimum. This spacecraft had walls made of foil material that was so thin, that a screwdriver could puncture it. Armstrong flew the lunar module while Aldrin called out flight data from the instruments.

It was a smooth flight until the *Eagle* began its descent. A series of program alarms gave flight controllers some anxious moments. The computer was overloaded with data and sounded the alarms.

*And I think everybody shares that observation, and I don't know why you have that impression, but it's so small, it's very colorful—you know, you see an ocean and gaseous layer, a little bit, just a tiny bit, of atmosphere around it, and compared with all the other celestial objects, which in many cases are much more massive, more terrifying, it just looks like it couldn't put up a very good defense against a celestial onslaught.*

— *Neil Armstrong (Discussing the Earth)*

Steve Bales, a twenty-six year old controller quickly analyzed the data and decided that the mission could proceed safely. He said, stuttering slightly as adrenaline rushed through his body, "We...we're go on that flight."

Neil Armstrong remained calm, as he did in other urgent situations he faced as a pilot and astronaut. The spacecraft was not behaving erratically and seemed stable, so Armstrong was not particularly worried.

During a training exercise in Houston, Armstrong was almost killed when his lunar lander training vehicle (LLTV) failed and exploded in a fireball. He narrowly escaped by ejecting, and reached the ground by parachute, suffering only a tongue laceration. Back on *Eagle*, another potentially serious problem did bother him.

> *Trying to get into a pretty tight spot probably wouldn't be fun. Also, the area was coming up quickly, and it soon became obvious that I could not stop short enough to find a safe landing spot...*
>
> — Neil Armstrong

With only seconds of fuel remaining, Armstrong overrode the computer and made a very soft landing on a smooth part of the Sea of Tranquility.

At the moment *Eagle's* probe made contact with the lunar surface, Aldrin stated, "Contact light." This was the very first broadcast from the lunar surface.

As commander, Armstrong stated, "Houston, Tranquility Base here. The *Eagle* has landed."

A voice from mission control responded, "Roger Tranquility. We copy you on the ground. You got a bunch of guys about to turn blue, and we're breathing again!"

## The Untold Story of the Apollo 11 Silicon Disc

*Armstrong training in the Lunar Lander Training Vehicle (LLTV) just weeks before the Moon launch. A vehicle similar to this exploded in a fireball just after Armstrong ejected from it.*

*LLTV Pilot Bud Ream congratulates Armstrong after a successful training flight. The LLTV made NASA leaders quite anxious since it was a dangerous flying machine — an aircraft engine with an astronaut strapped on top.*

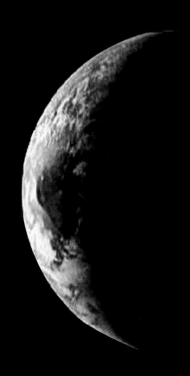

*Crescent Earth.*

*Fortunately, there were no really harrowing parts of the flight. The most difficult part, from my perspective, and the one that gave me the most pause, was the final descent to landing. That was far and away the most complex part of the flight. The systems were very heavily loaded at that time. The unknowns were rampant. The systems in this mode had only been tested on Earth and never in the real environment. There were just a thousand things to worry about in the final descent. It was hardest for the system and it was hardest for the crews to complete that part of the flight successfully.*

— *Neil Armstrong*

*I was absolutely dumbfounded when I shut the rocket engine off and the particles that were going out radially from the bottom of the engine fell all the way out over the horizon, and when I shut the engine off, they just raced out over the horizon and instantaneously disappeared, you know, just like it had been shut off for a week. That was remarkable. I'd never seen that. I'd never seen anything like that. And logic says, yes, that's the way it ought to be there, but I hadn't thought about it and I was surprised.*

— *Neil Armstrong*

# The wine curled slowly and gracefully

The Eagle had landed on the untouched surface of a stark and barren lunar landscape. The sky had a velvet black appearance that was much blacker than any night on Earth. The Sun's brightness made it feel like an empty lit up football stadium at night. Inside the spacecraft, Armstrong and Aldrin paused to look at each other and briefly shook hands. No specific ceremony was planned at this point, except for a private one.

Buzz Aldrin and his pastor had planned a religious ceremony, in the form of a Presbyterian Holy Communion. Due to the legal and political doctrine which separated church and state, NASA wanted this to be a private affair for Aldrin. During Apollo 8 the astronauts read from the Bible, which offended some groups. NASA wanted to avoid that problem again.

Buzz stated, "Houston, this is the LM pilot speaking. I would like to request a few moments of silence. I would like to invite each person listening in, wherever or whoever he may be, to contemplate the events of the last few hours and to give thanks in his own individual way." He read from a note (see illustrations) he took with him which stated, "My way shall be by partaking of the elements of Holy Communion." This portion was not stated aloud on radio. Next, he quietly read to himself:

"An Jesus said, I am the vine, you are the branches. Whoever remains in me, and I in him, will bear much fruit; for you can do nothing without me."

He also read Psalms 8:

"When I consider thy heavens, the work of thy fingers, the moon and the stars, which thou has ordained; what is man, that thou are mindful of him? And the Son of Man, that thou visitest Him?"

Aldrin took along a small silver chalice, the Host (bread) and a tiny vial of wine. After the landing, he poured the wine into the chalice. Aldrin stated that the wine, "Looked like syrup as it swirled around the sides of the cup in the light gravity, before it settled at the bottom."

Houston This is Eagle
The LM Pilot speaking
I would like to request a few moments of silence over, to invite
I would like for each person listening in, whereve and whomever he may be, to contemplate for a moment the events of the past few hours and to give thanks in his own individual way. —

My way shall be by partaking of the elements of Holy Communion

*Actual images of handwritten notes by Buzz Aldrin, flown to the Moon on Apollo*
*— Courtesy of Buzz Aldrin and Heritage Auctions*

As Jesus said.

"I am the vine, you are the branches. Whoever remains in me, and I in him, will bear much fruit; for you can do nothing without me."

Psalms 8:3,4

"When I consider thy heavens, the work of thy fingers, the moon and the stars, which thou has ordained; What is man, that thou art mindful of him? and the Son of Man, that thou visitest Him?"

> *In the one-sixth gravity of the Moon, the wine curled slowly and gracefully up the side of the cup. It was interesting to think that the very first liquid ever poured on the Moon, and the first food eaten there, were Communion elements. Just before I partook of the elements, I read the words which I had chosen to indicate our trust: John 15:5, "I am the vine, you are the branches. He who abides in me, and I in him, he it is that bears much fruit."*
>
> — Buzz Aldrin

Viewers on Earth celebrated a moment of silence while he partook in Communion. Most people were not aware of this ceremonial event. The silicon disc also has Psalms 8 in the message from the Vatican.

The astronauts put on their life support systems and the lunar module was depressurized. Armstrong began his descent from the ladder and manually triggered a small television camera mounted on the outside of *Eagle* that captured the moment. He got to the bottom of the ladder and jumped down to the lunar module pad.

As Armstrong stepped off the pad, he stated with great eloquence and humility, "That's one small step for a man, one giant leap for mankind."

Neil Armstrong was the only person who knew what he would say. These famous words were not part of any NASA committee decision or directive. As commander of Apollo 11, he was given the authority to say whatever he felt was appropriate. Not one person, including his family or the other crew members, knew what he would say. He thought of this statement while he was still in the lunar module.

There has been much speculation about how and why he spoke these words. Shapley's earlier memorandum had stated that "A forward step for all mankind" was one of the objectives. Regardless of what influenced Armstrong's words, they are one of the most recognized phrases in history.

> *"That's one small step for (a) man, one giant leap for mankind."*
> — Neil Armstrong

There has been controversy surrounding the "a" in his statement. People on Earth heard him say, "That's one small step for man, one giant leap for mankind." Historians assumed that he bungled the words and Armstrong himself was unsure. He certainly knew what he meant to say. The missing "a" has recently been recovered through modern acoustical evidence. A computer analysis of Armstrong's transmission from the Moon was done by Peter Shann Ford, a Sydney, Australia-based computer programmer. It provides convincing proof that the "a" was hidden in some static during the transmission: "That's one small step for (a) man, one giant leap for mankind." James R. Hansen, Armstrong's authorized biographer, presented the findings to Neil Armstrong. Presentations were next made at the Smithsonian Institution's Air and Space Museum in Washington, D.C. and at NASA's Washington headquarters.

Armstrong's first order of business was supposed to be to obtain a contingency sample of the lunar soil. In case of an emergency, the astronauts might have to abort the mission and leave quickly. The contingency sample insured that at least a small amount of the lunar soil would be brought back to Earth. Instead, Armstrong, like any other tourist in an exciting new location, quickly began taking the first breathtaking pictures of the lunar surface. Mission Control kindly reminded him to get the sample. Aldrin joined him later and described the vast beauty of the Moon as "Magnificent desolation."

*Buzz Aldrin called the Moon's surface, "Magnificent desolation."*

*I was surprised by the apparent closeness of the horizon. I was surprised by the trajectory of dust that you kicked up with your boot, and I was surprised that even though logic would have told me that there shouldn't be any, there was no dust when you kicked. You never had a cloud of dust there. That's a product of having an atmosphere, and when you don't have an atmosphere, you don't have any clouds of dust.*

*— Neil Armstrong*

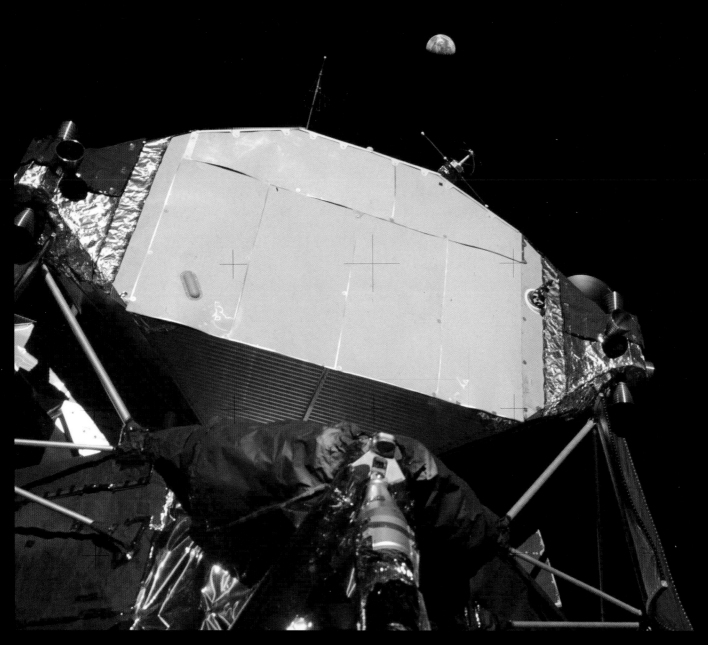

*The Earth was above the astronauts like a tiny blue marble.*

## The Untold Story of the Apollo 11 Silicon Disc

*Buzz Aldrin. Neil Armstrong's reflection can be seen in the visor along with a very tiny blue dot — the Earth's reflection.*

## Plaque unveiled and flag is (barely) planted

Neil Armstrong unveiled the plaque, which was covered and protected by a shield. Armstrong read the words on the plaque to the millions of people listening and watching on Earth. Next, they turned their attention to the planting of the American flag. This turned out to be a much more difficult problem than anticipated. The base of the pole was stuck only about six inches into the subsurface crust of the lunar soil. Together, the astronauts tried hard to extend the horizontal telescoping rod, but they could not pull it all the way out. Armstrong, an Eagle Scout, had dealt with many flags in his career. Now he and Aldrin feared that the flag might topple over. It was close to being a public relations debacle. Luckily, the flag stayed upright, but had a strange looking curl. The astronauts proudly saluted the flag and then received a most unusual phone call- from the White House:

*Hello, Neil and Buzz. I'm talking to you by telephone from the Oval Room at the White House and this certainly has to be the most historic phone call ever made. For one priceless moment in the whole history of man, all the people on this Earth are truly one.*
— *Richard Nixon*

*Plaque rests on the moon. Armstrong had climbed down this ladder earlier.*

*Buzz Aldrin.*

## Silicon disc is forgotten

Neil Armstrong and Buzz Aldrin completed their busy two and a half hour Moonwalk and deployed their lunar experiments. Armstrong even spent about three minutes "running" to explore a crater. His short adventure away from Eagle was a bit risky, as mission control would not have approved of him straying away too far. Due to the complexity of the mission flight plan, the astronauts nearly forgot about the disc. As Aldrin climbed up the ladder back into the spacecraft Eagle, Armstrong reminded Aldrin about the silicon disc and the other memorial items. (Note: The timestamps in the transcript below are relative to the time of launch—111 hours, then the minutes and seconds.)

*111:36:38 Armstrong: How about that package out of your sleeve? Get that?*

*111:36:53 Aldrin: No.*

*111:36:55 Armstrong: Okay, I'll get it. When I get up there (to the porch). (Pause)*

**Aldrin later recollected:**

> We had forgotten about this up to this point. And I don't think we really wanted to totally openly talk about what it was. So it was sort of guarded. And I knew what he (Armstrong) was talking about.
>
> — *Buzz Aldrin*

## Disc is tossed onto lunar surface

Next, Buzz Aldrin reached into his left shoulder pocket and removed the beta cloth pouch containing the disc and the memorial items. He tossed the package from *Eagle*. It fell next to Armstrong's right side, on the fine lunar dust of the Sea of Tranquility. Armstrong gave it a nudge, as if to acknowledge its presence and to remove some dust that had covered it. Secured in an aluminum case and a beta cloth pouch, and with only one sixth of the gravity of Earth, the thin, fragile silicon disc landed gently, without being damaged. For the first time since the presence of humans on the Earth, powerful and poetic messages would be left on an alien world. This bizarre world was one in which an unprotected human being would suffocate and his blood would boil within minutes. The main objectives of the mission, like collecting some lunar rocks, had already been completed. With precious time expiring, picking up the disc would have been awkward for Armstrong in his bulky suit and stiff gloves. No ceremony or other special statements were made. President Nixon and the astronauts had done this earlier with the unveiling of the plaque attached to the lunar module and the planting of the American flag.

> 111:37:02 Aldrin: Want it now?
>
> 111:37:06 Armstrong: Guess so. (Pause)
>
> 111:37:15 Armstrong: (Asking Buzz Aldrin if he likes the placement of the package) Okay?
>
> 111:37:15 Aldrin: Okay. (Pause)

*Then, later,*

*114:52:18 Garriott: Tranquility Base, Houston. Over. (Pause)*

*114:52:27 Armstrong: Go ahead, Houston.*

*114:52:28 Garriott: Roger. Two more verifications, here. Will you verify that the disk with messages was placed on the surface as planned, and also that the items listed in the flight plan — all of those listed there — were jettisoned? Over.*

*114:52:48 Armstrong: All that's verified.*

*114:52:51 Garriott: Roger. Thank you, and I hope this will be a final good night.*

*114:52:57 Armstrong: Okay.*

We were so busy that I was halfway up the ladder before Neil asked me if I had remembered to leave the mementos we had brought along. I had completely forgotten. What we had hoped to make into a brief ceremony, had there been time, ended almost as an afterthought. I reached into my shoulder pocket, pulled the packet out and tossed it onto the surface.

— *Buzz Aldrin*

*Buzz Aldrin descends the ladder to join Neil Armstrong*

# The Untold Story of the Apollo 11 Silicon Disc

*Buzz Aldrin.*

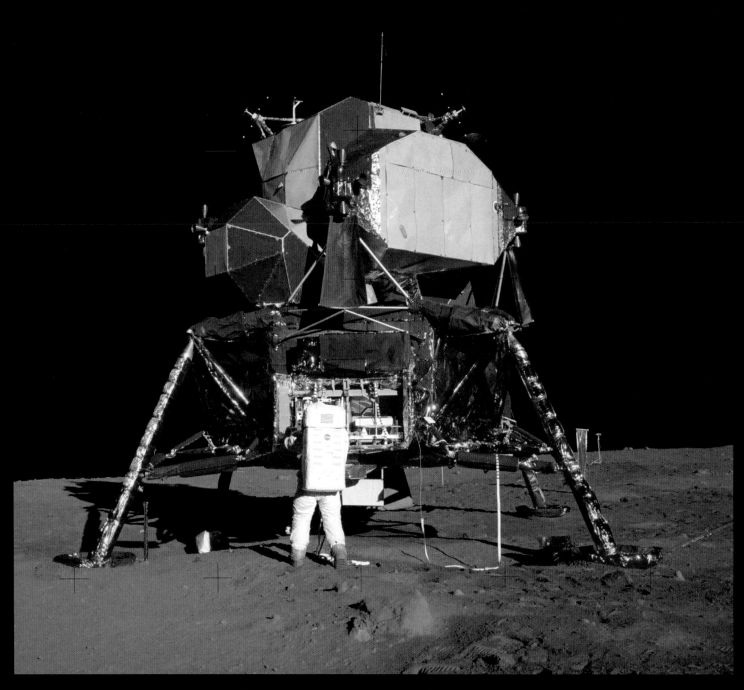

*Buzz Aldrin removing experiments from Lunar Module.*

*Buzz Aldrin next to the Lunar Seismograph used to detect moon quakes.*

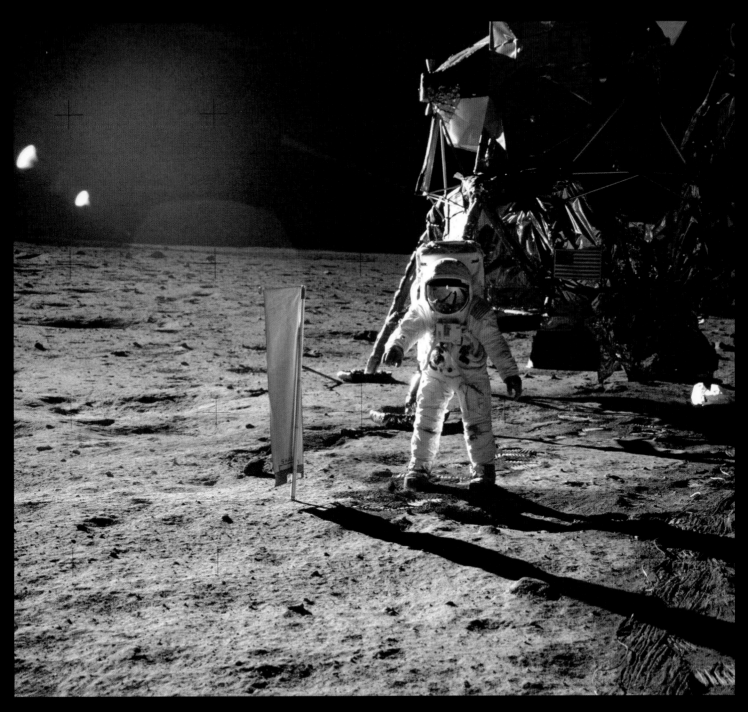

*Buzz Aldrin next to the solar wind experiment.*

*Buzz Aldrin pointed his camera down to show his boot print on the fine lunar dust.*

> *Human character—this is the area where we've made the least progress—learning more about the brain, about our behavior and the ways we relate to one another. I think that's the most important direction we can take in the next twenty years; basically to begin to understand ourselves.*
>
> — Neil Armstrong

The disc was obviously not the most important aspect of the mission. Since it had been added to their mission as an afterthought, placing it on the Moon was not practiced much during simulations. Planting the U.S. flag, unveiling the plaque, and the phone call from Nixon were not rehearsed, either. The astronauts' primary concern was how to survive in a hostile lunar environment; the very elements of temperature and radiation extremes that engineers designed the spacecraft, plaque, flag, and the tiny silicon disc to withstand. They climbed back into Eagle with about 21 kilograms of Moon rocks. The piece of the Wright flyer remained in the spacecraft throughout the Moonwalk.

The astronauts tried to get some sleep during the mission. Armstrong was bothered by the Earthshine that beamed through the telescope used by the astronauts for navigation.

Galileo Galilei, an Italian astronomer, was the first person to describe the mountains and craters of the Moon that he observed through his telescope almost 300 years earlier. His ideas were considered heretical and condemned by the Church. Psalm 104:5 says, "(the Lord) set the Earth on its foundations; it can never be moved." Galileo was forced to recant his belief that the Sun was the center of the solar system (heliocentric).

Now, Pope Paul VI sent an ornate, framed message from Psalm 8, and a handwritten and signed note, all etched onto the silicon disc — a message that could outlast even the beauty of Michelangelo's Sistine Chapel:

> *To the glory of the name of God who gives such power to men, we ardently pray for this wonderful beginning.*
>
> — *Pope Paul VI, etched on Moon disc*

The lunar dust, like ashes from a fireplace, covered the astronauts. The pungent smell of spent gunpowder filled the air in the small pressurized spacecraft. Armstrong and Aldrin left behind the descent stage of the lunar module, which served as a base for the ascent stage. The descent stage read "United States" and had a flag decal. The plaque remained attached to it. A handwritten note on a NASA memorandum discussing the base decals jokingly stated, "You mean no 'Kilroy was here?'" The reference was to an American popular culture saying.

The ascent stage of *Eagle* lifted the two astronauts off the Moon to lunar orbit. They completed a rendezvous and docked with *Columbia*, still piloted by a confident Collins. The precious cargo of Moon rocks was transferred into *Columbia* along with the two Moonwalkers. The trio of astronauts closed the hatch between the two spacecraft and cut the *Eagle* loose. It slowly drifted away and eventually impacted into the Moon like the plethora of meteorites that have struck its surface for billions of years.

The flag that Armstrong and Aldrin had desperately planted on the moon was abruptly toppled by the ascent engine thrust. A myriad of footprints and discarded material remained behind. The prints would last longer than any other human footprints in a strange world that has no erosion.

The Moon slowly shrank as the three astronauts began the voyage home. The crew returned to Earth safely, and the national anthem was finally played, with a joyous President Nixon attending the brief ceremony. The astronauts remained in a quarantine facility for several weeks. This was done to insure that the astronauts did not bring back "Moon germs."

*Eagle on the way home.*
*The Descent stage remained on the surface of the Moon with the plaque still attached to it.*

*Nixon jokes with the astronauts in their isolation chamber.*

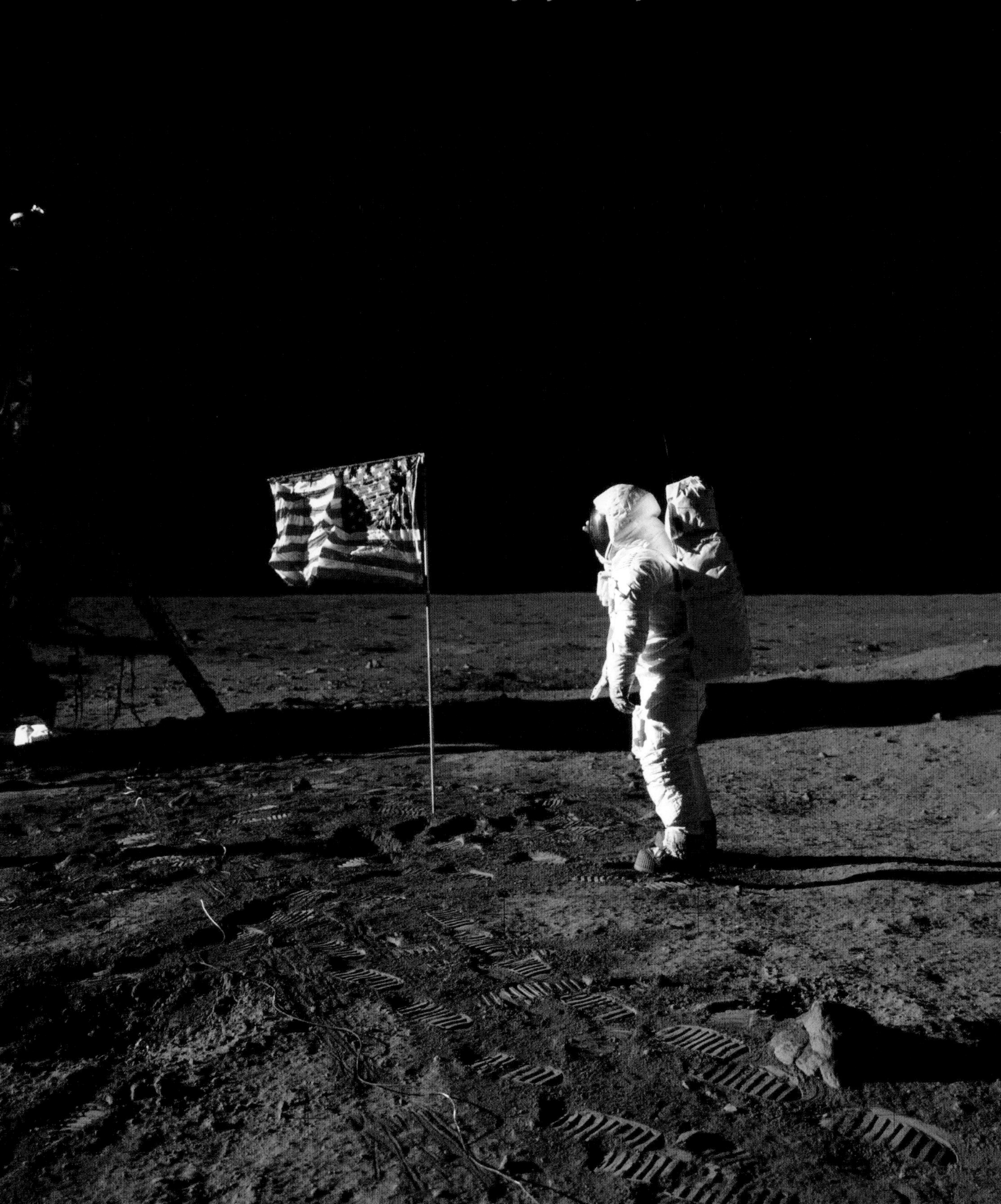

# De revolutionibus orbium coelestium
## The revolutions of the heavenly spheres

 Getting to the moon would require the work and knowledge of many scientific predecessors including: Nicolaus Copernicus, Johannes Kepler, and Isaac Newton.

 The Earth's intricate motions of daily rotation, annual revolution, and the tilting of the axis were written in Copernicus' De revolutionibus orbium coelestium (1543). He also accurately described the order of the planets and the motions of the Sun and Moon.

 Later, Isaac Newton described his three famous laws of motion in Philosophiæ Naturalis Principia Mathematica (1687). One such law was an astonishing scientific proof that an object in motion would continue moving until it was made to stop by another force. Since there is no wind resistance in space, once a spacecraft was headed for the Moon, it would keep going until it got there, without expenditure of fuel. The velocity needed to escape the Earth's gravity and achieve Earth orbit was also something tangible. The foundations of math and physics would make a voyage to the Moon a theoretical possibility.

# The Untold Story of the Apollo 11 Silicon Disc

*If I had seen further, it is by standing on the shoulders of giants.*
— *Sir Isaac Newton, in a letter to Robert Hooke, 1675*

*Flag seen from window of the Lunar Module amidst a myriad of foot prints.*

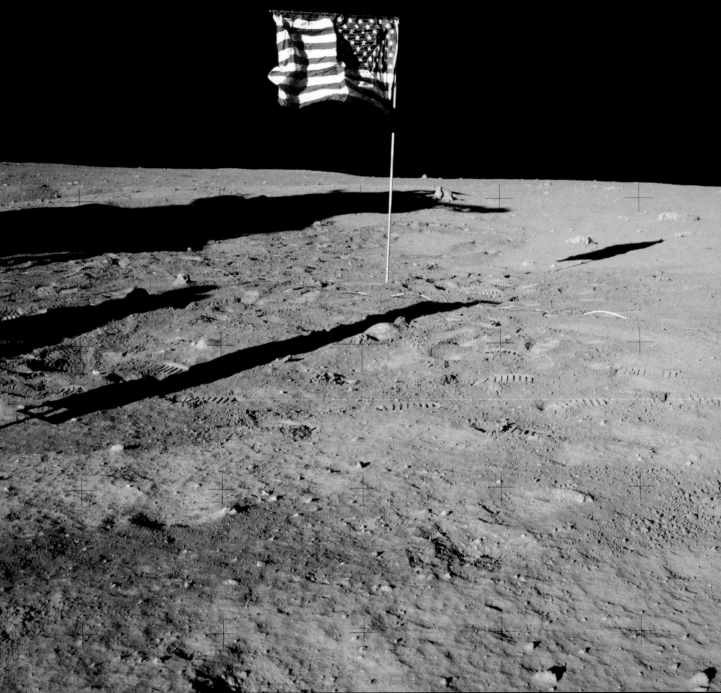

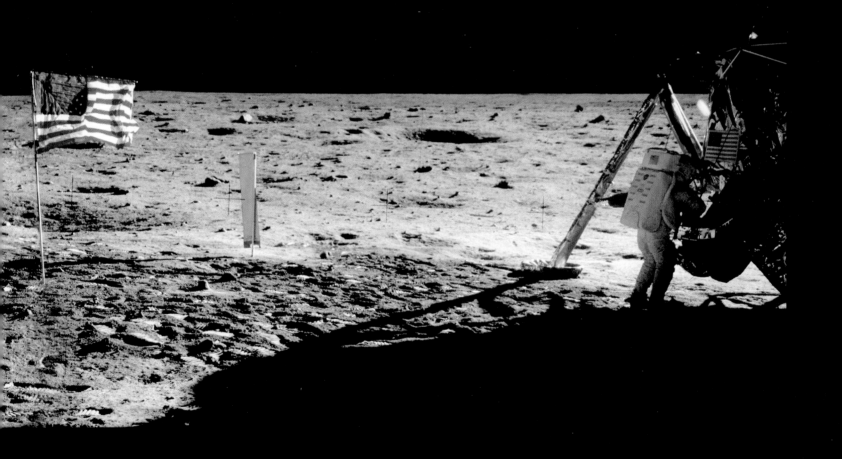

*One oversight of the mission was forgetting to photograph Neil Armstrong on the lunar surface. Buzz Aldrin captured this panoramic image of Neil Armstrong.*
— *Courtesy of Moonpans.com*

The Apollo 11 crew received a ticker tape parade in New York City. More people attended than at the end of World War II, or for Charles Lindbergh's flight.

The astronauts completed a "Giant Step" tour of the world, which included visits to Mexico City, Bogota, Buenos Aires, Rio de Janeiro, Grand Canary Island, Madrid, Paris, Amsterdam, Brussels, Oslo, Cologne, Berlin, London, Rome, Belgrade, Ankara, Kinshasa (Congo), Tehran, Bombay, Dacca, Bangkok, Darwin (Australia), Sydney, Guam, Seoul, Tokyo, Honolulu, and Houston. Neil Armstrong later visited the Soviet Union. He also joined Bob Hope's 1969 tour to entertain troops in Vietnam. They stopped in Germany, Italy, Turkey, Taiwan, and Guam.

The pieces of the Wright flyer were returned to the Earth. They had flown above the sands of the beach at Kitty Hawk, North Carolina and now had voyaged to the sands of the Sea of Tranquility on the Moon. Armstrong was particularly proud of this item.

# Armalcolite

America had reached the pinnacle of world technological advancement. People would soon use the phrase, "They can get a man on the Moon, but they can't____."

The Moon rocks appear gray and dark, and are proudly displayed in museums for people to admire. Armstrong and Aldrin collected 48 pounds (21 kg) of lunar soil and rocks. The samples Armstrong collected were among "the best" of any of the lunar landings, according to Apollo 17 Moonwalker and geologist Harrison Schmitt. Only ten percent of the Moon rocks have been studied. Ninety percent are for prosperity and future scientific investigation.

The Moon rocks consist of anorthrosite, breccia, and basalt. There are also some beautiful glass particles, some green or orange. The astronauts had a mineral named after them, called Armalcolite. It has the following chemical formula: $(Mg,Fe^{++})Ti_2O_5$. The name is taken from the Apollo 11 astronauts: ARMstrong, ALdrin, and COLlins. The mineral was found at Tranquility Base. It is also naturally present on Earth.

# The Moon - a viable energy source

Compared to Earth, a higher quantity of Helium-3 exists on the Moon. Helium-3 is a gas produced by the Sun. The Earth's atmosphere blocks solar flares that contain helium-3. Since the Moon has no atmosphere, it has absorbed helium-3 for billions of years. Scientists, notably Apollo 17 astronaut Harrison Schmitt, believe that the energy contained would be economically viable for mining the helium-3. Scientists have known for decades how to make energy from deuterium-helium-3 fusion reactors. Further research is necessary to fund the engineering of such a reactor for practical use.

Schmitt's book, entitled *Return to the Moon*, details the way a private company could invest in a lunar mining operation. The book's Foreword, written by the first Moonwalker, states:

> *If you believe the Earth's increasing appetite for energy and the suspected future decrease in available energy will create an ever more severe problem for our Earth's future, you will find this proposal worthy of careful examination.*
>
> *— Neil Armstrong*

# *Vengeance*

World War I was supposed to be the "War to end all wars." Instead, it opened festering wounds of German pride. Annihilated and humiliated by the First World War, conditions were ripe for the Nazi party to form. That party and its predecessors rose because of a charismatic leader, Adolf Hitler.

Hitler blamed Jews and the allies for Germany's economic demise and becoming an impotent nation. A 1938 Time Magazine "Man of the Year" recipient, Hitler revived Germany's economy with an incredible armada of military might and scientific and architectural advancement, but he also began a campaign of hatred.

Six million Jews, and up to fourteen million people, were systematically gassed, starved, and executed by a new ideology. He was a cruel dictator with genocide on his list of things to accomplish.

America remained silent at the start of World War II. Franklin Roosevelt began a clandestine relationship with Winston Churchill to fight the Germans.

In the midst of the rise of German nationalism came a wave of unprecedented scientific advancement in the area of rocket design. Wernher von Braun, a gifted rocket scientist, created and manufactured some of the deadliest and feared machines of the war. Influenced by American rocket scientist Robert Goddard and Germany's Hermann Oberth, von Braun dreamed of space travel. Those dreams would remain unfulfilled when an evil fervor resulted in the development of rockets with only one intent—to destroy people.

The focal point was a program located in Peenemunde, in northern Germany. There, a weapon called the V-2 rocket was developed with the use of slave labor. The production of the weapon killed more people than its use. A sophisticated weapon, it stood for "vengeance," and it pulverized Great Britain. Londoners spent many nights in underground bunkers as the bombs exploded. The world stood in awe and fear as the Nazi military wiped out millions in Europe.

**The rocket worked perfectly except for landing on the wrong planet.**
— *Wernher von Braun, September 1944*

After the brutal Nazi regime finally came to an end, Von Braun's team of rocket scientists became a highly sought-after group by both the United States and the Soviet Union. Operation paperclip was the code name of the U.S. intelligence plan to find high-level German scientists for American interests.

Von Braun intentionally surrendered to the Americans, as he feared that the Soviet Union was run by a ruthless, totalitarian leadership. As an American, von Braun wanted be in charge of America's space program. The American President, Dwight D. Eisenhower, had some difficult decisions to make. Letting a former Nazi run the show was not exactly a popular public relations strategy. However, the launch of Sputnik by the Soviet Union quickly changed priorities.

*President John F. Kennedy and John Glenn. Glenn was the first American to orbit the Earth on February 20, 1962.*

# The Space Race

The successful launch of the Soviet Sputnik satellite in 1957 marked the beginning of a competition between the superpowers for dominance in the space race. The tiny satellite weighed only 183 pounds, but was a startling development. The implication was that a nuclear warhead could be dropped on American cities. Sergey Pavlovich Korolyov was the chief Russian rocket designer who kept the Soviet Union ahead of the United States. His death in 1966 may have lead to the Soviet Union's inability to land a man on the Moon.

The Soviets would stun the world again with the first man in space. In 1961, cosmonaut Yuri Gagarin made an unprecedented orbital flight that clearly demonstrated Soviet pre-eminence in space. The Russians kept their failures private and touted their accomplishments with flamboyant propaganda.

American and Soviet national pride was now also at stake.

The United States had a slow start to its inexperienced space program. Von Braun, who was angered by America's passivity, was given a larger role in the space program. He was finally able to get a U.S. satellite called Explorer into Earth orbit. In 1959, seven Americans were chosen as the first astronauts: Malcolm S. Carpenter, Leroy G. Cooper, Jr., John H. Glenn. Jr., Virgil I. Grissom, Walter M. Schirra, Jr., Alan B. Shepard, Jr., and Donald K. Slayton.

In 1961, Alan B. Shepard, Jr. made a fifteen minute sub-orbital flight, making him the first American in space. This was nothing compared to the Soviet cosmonauts who made multiple orbits with each flight. They intricately maneuvered their spacecraft to do this. America was far behind, and needed a vision to succeed.

A youthful and dynamic President John F. Kennedy made an announcement that invigorated NASA and invoked a forceful and seemingly impossible message to Congress:

*I believe that this nation should commit itself to achieving the goal, before this decade is out, of landing a man on the Moon and returning him safely to the Earth. No single space project . . . will be more exciting, or more impressive to mankind, or more important . . . and none will be so difficult or expensive to accomplish . . .*

— *President John F. Kennedy, May 1961 (message etched onto silicon disc)*

*Buzz Aldrin inside Eagle.*

# The Untold Story of the Apollo 11 Silicon Disc

*Neil Armstrong seen with the lunar surface in the window behind him.*

*A smiling Neil Armstrong in the Lunar Module while on the moon.*

# The Untold Story of the Apollo 11 Silicon Disc

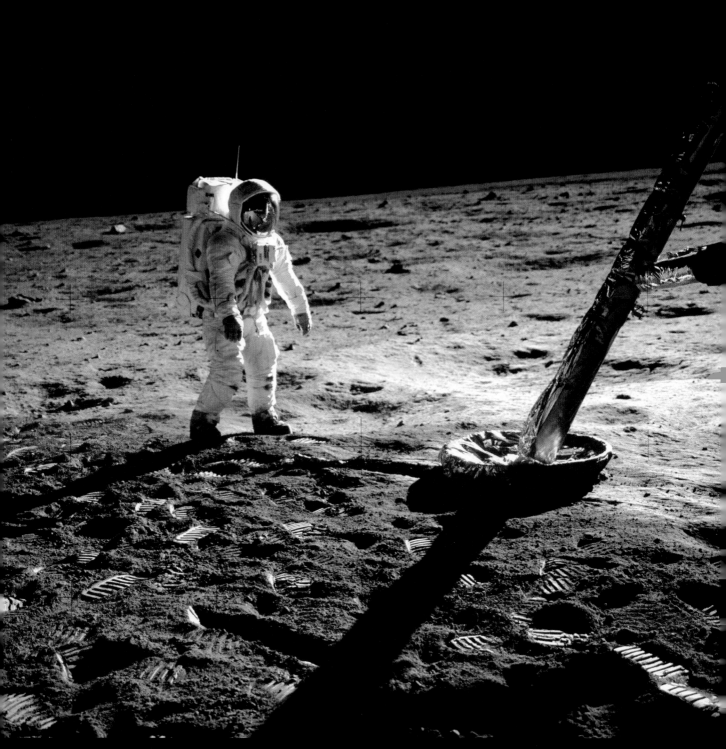

*Buzz Aldrin standing next to a strut of the Lunar Module*

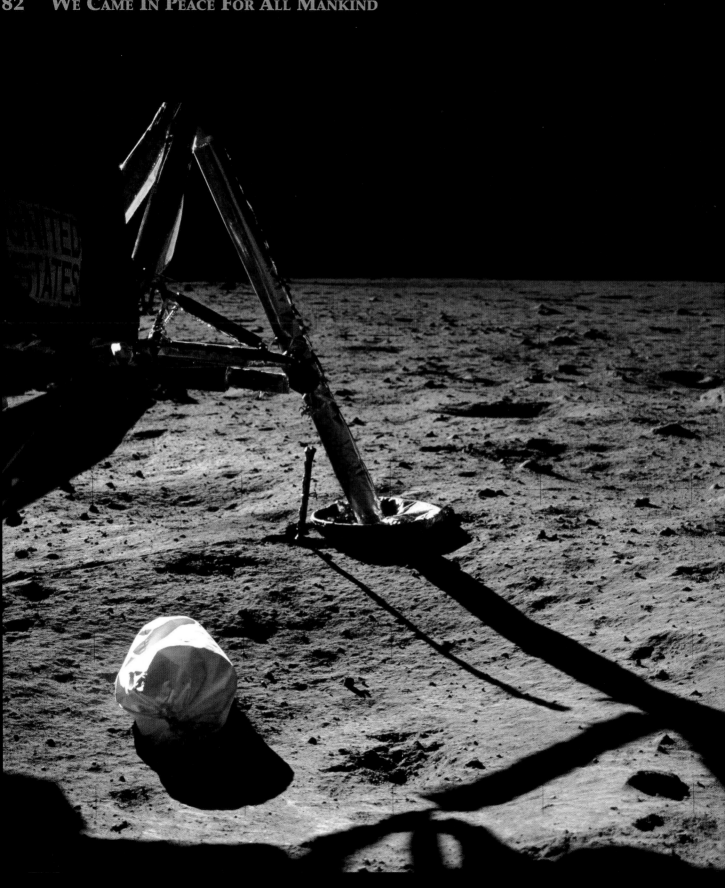

To honor the father of modern rocketry, Buzz Aldrin took two of Robert Goddard's books to the Moon. He presented one of them to Goddard's widow on return from the Moon. Goddard's foundation of rocket science led to the Saturn V rocket. The majestic Saturn V rocket had three stages. The cluster of five F-1 engines produced 1.5 million pounds of thrust each. The rocket had over 1 million parts and was the most successful rocket ever created. Once the lunar module docked with the command module in space, the joint spacecraft traveled to the Moon at over 24,000 miles per hour. NASA plans to use a similar rocket design to return to the Moon.

# Moon dust around the world

In November 1969, President Nixon requested that NASA create approximately 250 displays for distribution by the White House. They contained lunar surface material and the flags of 135 nations and U. S. states and territories.

Each presentation included 0.05 grams of lunar dust (encased in Lucite) retrieved by the Apollo 11 astronauts, as well as a flag of the recipient nation carried aboard the mission.

The displays presented to foreign nations were inscribed:

"Presented to the People of ___ by Richard Nixon, President of the United States of America. This Flag of your nation was carried to the Moon and back by Apollo 11, and this fragment of the Moon's surface was brought to Earth by the crew of that first manned lunar landing."

The following nations received a display:

| | | | |
|---|---|---|---|
| Afghanistan | Botswana | Chile | Dominican Republic |
| Albania | Brazil | China | Ecuador |
| Algeria | Bulgaria | Columbia | El Salvador |
| Andorra | Burma | Congo/Brazzaville | Equatorial Guinea |
| Argentina | Burundi | Congo (Kinshasa) | Ethiopia |
| Australia | Cambodia | Costa Rica | Finland |
| Austria | Cameroon | Cuba | France |
| Barbados | Canada | Cyprus | Gabon |
| Belgium | Central African Rep | Czechoslovakia | Gambia |
| Bhutan | Ceylon | Dahomey | Germany |
| Bolivia | Chad | Denmark | Ghana |

| | | | |
|---|---|---|---|
| Greece | Lebanon | Nicaragua | Sudan |
| Guatemala | Lesotho | Niger | Swaziland |
| Guinea | Liberia | Nigeria | Sweden |
| Guyana | Libya | Norway | Switzerland |
| Haiti | Liechtenstein | Pakistan | Syria |
| Honduras | Luxembourg | Panama | Tanzania |
| Hungary | Madagascar | Paraguay | Thailand |
| Iceland | Malawi | Peru | Togo |
| India | Malaysia | Philippines | Trinidad and Tobago |
| Indonesia | Maldive Islands | Poland | Tunisia |
| Iran | Mali | Portugal | Turkey |
| Iraq | Malta | Romania | Uganda |
| Ireland | Mauritania | Rwanda | United Arab Rep. |
| Israel | Mauritius | San Marino | United Kingdom |
| Italy | Mexico | Saudi Arabia | Upper Volta |
| Ivory Coast | Monaco | Senegal | Uruguay |
| Jamaica | Mongolia | Sierra Leone | Venezuela* |
| Japan | Morocco | Singapore | Vietnam |
| Jordan | Muscat and Oman | Somalia | Western Samoa |
| Kenya | Nauru | South Africa | Yemen |
| Korea | Nepal | Southern Yemen | Yugoslavia |
| Kuwait | Netherlands | Soviet Union | Zambia |
| Laos | New Zealand | Spain | |

Displays were also presented to the United Nations and the Vatican.

*The flag for Venezuela was discovered to have been left off the Apollo 11 mission. Instead, a flag flown on Apollo 12 was used with the following wording: "This flag of your nation was carried to the Moon and back, and this fragment of the Moon's surface was brought to earth by the crew of the first manned lunar landing."

## We Came in Peace

Today, on the lunar surface at the Sea of Tranquility, a few objects remain from Apollo 11; the base of the lunar module and its plaque, some memorials, a few scientific instruments, the American Flag, and the silicon disc.

One unintentional aspect of the lunar landing was to unite the world. We learned that manned exploration of the unknown did something special to our hearts and minds. It refocused our differences.

Perhaps most symbolic of the lunar memorial items is the silicon disc, etched with the hopes of mankind for peace, and a spirit of celebration of our most important adventure.

> May God grant the skill and courage which have enabled man to alight upon the Moon will enable him, also, to secure peace and happiness upon the Earth and avoid the danger of self-destruction.
>
> — Eamon de Valera, message from Ireland etched onto silicon disc

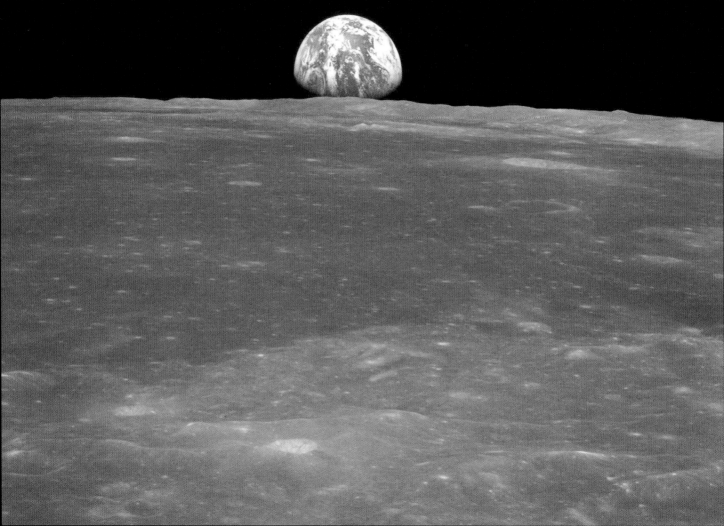

# Beyond the broken wings of Icarus

NASA implemented three programs, Mercury, Gemini, and Apollo. The Mercury missions focused on Earth orbital flights. John Glenn made the first American Earth orbital flight.

The Gemini phase perfected rendezvous, docking, and endurance missions. During Gemini 8, Neil Armstrong and Dave Scott made the first successful docking with an unmanned Agena spacecraft. A stuck thruster put Armstrong and Scott into a life-threatening tumble. The astronauts aborted the mission and survived. Armstrong proved he could be calm in perilous events.

In Gemini 12, the brilliant M.I.T. graduate, Buzz Aldrin performed a flawless space walk for over two hours which proved that astronauts could work on the outside of spacecraft safely and effectively.

The Apollo program would make Moon voyages. Apollo 8 made the first lunar orbital flight—a daring flight with no backup plan in case of failure. On Christmas Eve, 1968, James Lovell, Bill Anders, and Frank Borman made ten orbits around the Moon. America had beaten the Russians to the Moon.

Only one step remained: to land on its surface.

Apollos 9-10 practiced lunar and command module maneuvers in Earth and lunar orbit. The Moon landing was now very close to becoming a reality.

I am particularly proud speaking on behalf of the Greek nation, whose ancestors had the privilege to the forerunners in philosophical thought and scientific research, which first penetrated the universe.

It is a happy coincidence that the amazing program of man's flight to space, which has been so magnificently fulfilled today, bears the name of the Greek God Apollo; this symbolic name demonstrates the never ending effort of man to achieve knowledge, beyond time and place.

The difficulties, which had once broken the wings of Ikarus, are surpassed by man's persistence in his search for truth, and he is staring from the Moon at the Earth whose peace and welfare should be now, more than ever, his main preoccupation.
— *George Zoitakis, message from Greece etched onto silicon disc*

Silicon is also the basis of our new century. The technological revolution that we have become dependent upon is rooted in silicon-based microchips.

As many of the messages from the world warn, a lack of human connectivity and tolerance for each other can disturb its unity. The Apollo program reminded us that our Earth is the only place we have to live and that we must take care of it. Resting on the Moon, the silicon disc would survive any calamity on Earth. Only a rare meteor impact could jar it from its resting place.

The human species is complex, its character often questionable and flawed. *In Carrying the Fire*, a book by astronaut Michael Collins, he wrote about the lack of boundaries between countries when viewed from a quarter of a million miles in space. When viewed from the Moon, mankind lives on a fragile, marble-sized Earth, which can be covered with just a fingertip. Collins wondered how people might feel about their lives and conflicts if they could see what he and a few other courageous astronauts saw during the Apollo program.

*For, in the final analysis, our most basic common link is that we all inhabit this small planet. We all breathe the same air. We all cherish our children's future. And we are all mortal.*

*— John F. Kennedy*

*We will see a manned scientific base being built on the Moon. It'll be a scientific station manned by an international crew, very much like the Antarctic station. But there is a much more important question than whether man will be able to live on the Moon. We have to ask ourselves whether man will be able to live together down here on Earth.*

— *Neil Armstrong*

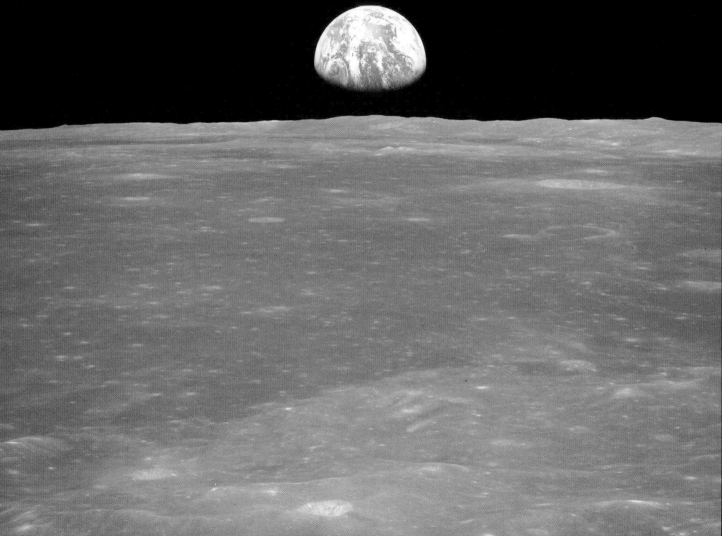

*This early version of the disc has the Vatican's image as a rectangle in the center. The lunar plaque with the Earth's hemispheres is visible at the top. The words "Apollo 11 Goodwill messages" is also etched on the disc. The four messages at the bottom are the U.S. Presidential Statements as well as the members of Congress and the leaders of NASA.*

# The Library of Goodwill Messages

The following pages contain the 74 messages etched onto the silicon disc (73 from foreign nations and one from U.S. presidents). There is also a miniaturized version of the lunar plaque at the top of the disc. The bottom portion contains the U.S. presidential seal and has excerpts of speeches from Eisenhower, Kennedy, Johnson and Nixon. The members of both houses of Congress and NASA administrators are also present.

The original paper messages (now in the Library of Congress) were used in this book where possible. They are the best images for reproduction. Viewing the messages through a microscope was quite exhilarating. However, photographing the images through a microscope posed several challenges. Lighting conditions and the size of the field of view were limitations. A Nikon *cool pics* digital camera and flexible epi-lighting were used to illuminate the disc from above. The disc is just 3.8 cm (1.5 inches) in diameter and only 14/1000th of an inch thick. Several messages were not available from the Library of Congress collection. In those cases, the microscope images were used instead. Those images are occasionally blurry outside the center of the microscope's field of view.

The images on the disc were in random order. They are placed in alphabetical order here for convenience. Appendix one has short vignettes of each leader. Responses from other nations are included after this library of messages.

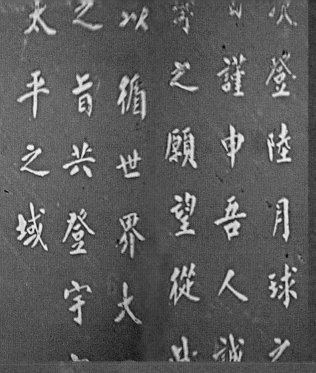

## The Netherlands

Ik heb grote bewondering voor de vaardigheid
en de volharding van allen, die ertoe hebben bij-
gedragen om de eerste bemande vlucht naar de maan
mogelijk te maken. Ik hoop, dat deze prestatie
een zegen zal blijken te zijn voor de toekomst
van de gehele mensheid.

Juliana R.

I have great admiration for the skill and
perseverance of all those who have contributed
to make the first manned flight to the moon
possible. I hope that this achievement will
prove of great benefit for the future of
mankind.

### El Presidente de la República de Costa Rica

Me uno al deseo ferviente de todos los costarricenses por el
éxito de la histórica hazaña que se propone realizar la nave Apolo 11,
en cuanto ella representa el adelanto científico y técnico alcanzado por
el hombre en su lucha pacífica por conquistar los espacios y en cuanto los tripulantes de esa nave representan el valor, la voluntad, el espíritu de aventura y el ingenio humanos.

El enorme esfuerzo científico y técnico desplegado para llevar a la luna a los primeros hombres merece la gratitud de la humanidad porque de ese esfuerzo se derivarán nuevos logros para aumentar
el bienestar de las familias humanas. Con fe esperamos mejores días
para toda la humanidad porque si al esfuerzo ya realizado se agregan
luego nuevos empeños en pro de la justicia y de la libertad, como corresponden al respeto que se debe a cada ser humano y en pro de una
difusión mayor del amor al prójimo, los cuales empeños podemos esperar que se hallen estimulados por el espíritu de humildad que habrá
de derivarse de la conciencia más clara y vívida sobre la pequeñez de
este planeta que nos sirve de hogar en los espacios siderales.

Como representante de la Nación Costarricense hago llegar
nuestro saludo a los héroes del Apolo 11 y a todos quienes han hecho posible que se realice su histórica hazaña.

Casa Presidencial, San José, a los treinta días del mes de junio de mil
novecientos sesenta y nueve.

*Silicon Disc with microscopic images behind it.
Photo signed by Buzz Aldrin*

# Afghanistan

Dear Sir:

With reference to your letter dated June 23, 1969, I have the honor to convey to you the message of His Majesty, the King of Afghanistan, to the Apollo 11 astronauts who are scheduled for the moon landing on July 16, 1969.

"The Afghan people join me in most warmly congratulating the American people, particularly the intrepid astronauts and all those who have played a role in this historic and incredible journey.

The Afghan people express the hope that the expanded knowledge man now has of his universe will be used wisely in the cause of peace on earth and for the betterment of the condition of mankind."

MOHAMMAD ZAHIR
KING OF AFGHANISTAN

Sincerely yours,

Mohammad Akbar
Charge d'Affaires

Mr. T. O. Paine
Administrator
National Aeronautics and Space Administration
Washington, D. C.

The Afghan people join me in most warmly congratulating the American people, particularly the intrepid astronauts and all those who have played a role in this historic and incredible journey.

The Afghan people express the hope that the expanded knowledge man now has of his universe will be used wisely in the cause of peace on earth and for the betterment of the condition of mankind.

Mohammad Zahir
King of Afghanistan

# Argentina

NON OFFICIAL TRANSLATION

From the President of the Argentine Republic, General Juan Carlos Onganía, to the President of the United States of America, Mr. Richard Nixon.

In the name of the Government and people of Argentina I have the honor to wish you the very best for the success of the great exploit that the brave crew of the Apollo XI is going to undertake. The effort and the risks that this extraordinary scientific enterprise demands open unlimited possibilities for the creative ability of the human spirit and it will constitute without any doubts an incentive that will commit all people in order to make a better and finer world. I present Mr. President my highest consideration to you.

Juan Carlos Onganía

From the President of the Argentine Republic, General Juan Carlos Ongaía, to the President of the United States of America, Mr. Richard Nixon.

In the name of the Government and people of Argentina I have the honor to wish you the very best for the success of the great exploit that the brave crew of the Apollo XI is going to undertake. The effort and the risks that this extraordinary scientific enterprise demands open unlimited possibilities for the creative ability of the human spirit and it will constitute without any doubts an incentive that will commit all people in order to make a better and finer world. I present, Mr. President, my highest consideration to you.

Juan Carlos Onganía
President

# Australia

**MESSAGE FROM AUSTRALIA**

AUSTRALIANS ARE PLEASED AND PROUD TO HAVE PLAYED A PART IN HELPING TO MAKE IT POSSIBLE FOR THE FIRST MAN FROM EARTH TO LAND ON THE MOON. THIS IS A DRAMATIC FULFILMENT OF MAN'S URGE TO GO "ALWAYS A LITTLE FURTHER"; TO EXPLORE AND KNOW THE FORMERLY UNKNOWN; TO STRIVE, TO SEEK, TO FIND, AND NOT TO YIELD. MAY THE HIGH COURAGE AND THE TECHNICAL GENIUS WHICH MADE THIS ACHIEVEMENT POSSIBLE BE SO USED IN THE FUTURE THAT MANKIND WILL LIVE IN A UNIVERSE IN WHICH PEACE, SELF EXPRESSION, AND THE CHANCE OF DANGEROUS ADVENTURE ARE AVAILABLE TO ALL.

JOHN GORTON

JULY 1969

Australians are pleased and proud to have played a part in helping to make it possible for the first man from earth to land on the moon. This is a dramatic fulfillment of man's urge to go 'always a little further'; to explore and know the formerly unknown; to strive, to seek, and to find, and not to yield. May the high courage and the technical genius which made this achievement possible be so used in the future that mankind will live in a universe in which peace, self expression, and the change of dangerous adventure are available to all.

John Gorton
Prime Minister

# Belgium

Now that, for the very first time, men will land on the moon, we consider this memorable event with wonder and respect.

We feel admiration and confidence toward all those who have cooperated in this performance, and especially towards the three courageous men who take with them our hopes, as well as those, from all nations, who were their forerunners or who will follow them in space.

With awe we consider the power with which man has been entrusted and the duties which devolve on him.

We are deeply conscious of our responsibility with respect to the tasks which may be open to us in the universe, but also to those which remain to be fulfilled on this earth, so to bring more justice and more happiness to mankind.

May God help us to realize with this new step in world history better understanding between nations and a closer brotherhood between men.

BAUDOUIN,
King of the Belgians.

Now that, for the very first time, men will land on the moon, we consider this memorable event with wonder and respect.

We feel admiration and confidence toward all those who have cooperated in this performance, and especially towards the three courageous men who take with them our hopes, as well as those, from all nations, who were their forerunners or who will follow them in space.

With awe we consider the power with which man has been entrusted and the duties which devolve on him.

We are deeply conscious of our responsibility with respect to the tasks which may be open to us in the universe, but also to those which remain to be fulfilled on this earth, so to bring more justice and more happiness to mankind.

May God help us to realize with this new step in world history better understanding between nations and a closer brotherhood between men.

Baudouin
King of the Belgians

# Brazil

"AO CONGRATULAR-ME COM O GOVÊRNO E O POVO DOS ESTADOS UNIDOS DA AMÉRICA PELO ACONTECIMENTO DO SÉCULO, FORMULO VOTOS A DEUS PARA QUE ESSA BRILHANTE REALIZAÇÃO DA CIÊNCIA ESTEJA SEMPRE A SERVIÇO DA PAZ E DA HUMANIDADE".

ARTHUR DA COSTA E SILVA
Presidente do Brasil

In rejoicing together with the government and the people of the United States of America for the event of the century, I pray God that this brilliant achievement of science remain always at the service of peace and of mankind.

Arthur Da Costa E Silva
President

# Canada

PRIME MINISTER · PREMIER MINISTRE

Man has reached out and touched the tranquil moon. Puisse ce haut fait permettre à l'homme de redécouvrir la terre et d'y trouver la paix.

*Pierre Elliott Trudeau*

**Man has reached out and touched the tranquil moon. Puisse ce haut fait permettre a l'homme de redecouvrir la terre et d'y trouver la paix. (May that high accomplishment allow man to rediscover the Earth and find peace.)**

**Pierre Elliott-Trudeau
Prime Minister**

# Chad

LET THIS IMPORTANT STEP IN THE COSMIC CONQUEST REMAIN FOR GENERATION TO COME A MARK OF AN EPOCH OF FELLOWSHIP AND OF UNIVERSAL PEACE STOP THESE ARE THE EARNEST WISHES OF THE CHADIAN PEOPLE AND OF ITS GOVERNEMENT STOP

FRANCOIS TOMBALBAYE
PRESIDENT OF THE REPUBLIC OF CHAD

# Apollo 11 Goodwill Messages

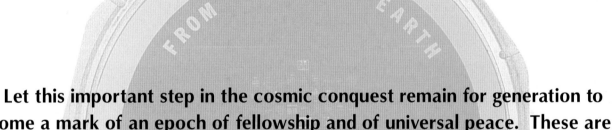

Let this important step in the cosmic conquest remain for generation to come a mark of an epoch of fellowship and of universal peace.  These are the earnest wishes of the Chadian people and of its government.

Francois Tombalbaye
President

# Chile

EMBAJADA DE CHILE

MENSAJE DEL PRESIDENTE DE CHILE, S.E. EDUARDO FREI PARA SER DEPOSITADO EN LA LUNA POR LOS ASTRONAUTAS DE APOLLO 11:

"Los hombres de nuestro planeta lleven hasta la luna un mensaje de paz y de buena voluntad de este lugar de la tierra que es Chile.

EDUARDO FREI,
Presidente de Chile"

May the men of our planet take to the moon a message of peace and good will from this place on the Earth that is Chile.

Eduardo Frei
President

# China

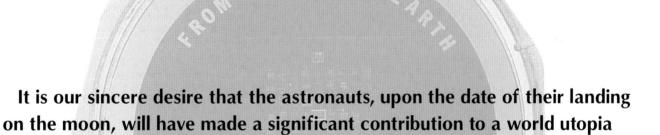

It is our sincere desire that the astronauts, upon the date of their landing on the moon, will have made a significant contribution to a world utopia and peace through the universe.

> Chiang Kai-Shek
> President, Republic of China

# Columbia

BOGOTA, D.E.
junio 27 de 1969

Señores Comandantes
Neil Armstrong,
Edwin Aldrin y
Michael Collins
Cabo Kennedy
Estados Unidos

        En el momento en que se preparan a cumplir una de las más extraordinarias hazañas de la historia les envío a nombre del Gobierno y del pueblo colombianos un fervoroso saludo con nuestros deseos de que élla sea coronada con un completo éxito.

        Quiero expresar también la admiración que todos los colombianos sienten hacia ustedes por su heroísmo personal; por los científicos y técnicos que han prestado el concurso de su conocimiento a esta magna empresa y por la gran nación norteamericana cuyo decidido apoyo ha hecho posible un proyecto que ayer no más parecía irrealizable.

        Les ruego dejar en la superficie de la luna con los demás objetos que recordarán la llegada del hombre por primera vez a nuestro satélite este mensaje que deseo tenga además el carácter de un símbolo de amistad entre los Estados Unidos y Colombia.

        Descenderán ustedes en la superficie lunar el día de nuestra fiesta nacional, precisamente en la fecha en que celebraremos el 159 aniversario de nuestra independencia.

        Los colombianos estaremos, pues, honrando la memoria de los próceres que cambiaron el rumbo de nuestro destino en la misma fecha en que ustedes estarán escribiendo una página inmortal en los anales de la humanidad.

As you prepare to undertake one of the most extraordinary feats in history, I wish to send you on behalf of the people and the Government of Columbia, a warm greeting with our wishes for the complete success of your mission. I also want to express the admiration of all Columbians for your personal heroism, for the scientists and technicians that have contributed their knowledge to this enterprise and for the great North American nation whose support has made possible a project that only yesterday appeared to be unfeasible.

Please leave on the moon along with the other objects that will bear witness of man's first arrival to our satellite, this message, as a symbol of friendship between Columbia and the United States.

You will descend upon the moon on our national holiday, when we observe the 159th anniversary of our independence. We, in Columbia, will be honoring the memory of the patriots that changed the course of our history on the same day when you will be writing an immortal page in the annals of mankind.

Carlos Lleras Restrepo
President

# Congo

The government of the Democratic Republic of the Congo follows with constant attention the achievements of human genius in the conquest of space in order to make man its master. The Congolese people, its party, its government, and myself express our ardent wish to see Apollo 11 successfully accomplish the mission which is our own. May these victories which have cost man so much energy and sacrifice contribute to the reinforcement of cooperation among peoples and serve peace for the greatest good of mankind. Best regards.

J.D. Mobutu
President

# Costa Rica

*El Presidente de la República de Costa Rica*

Me uno al deseo ferviente de todos los costarricenses por el éxito de la histórica hazaña que se propone realizar la nave Apolo 11, en cuanto ella representa el adelanto científico y técnico alcanzado por el hombre en su lucha pacífica por conquistar los espacios y en cuanto los tripulantes de esa nave representan el valor, la voluntad, el espíritu de aventura y el ingenio humanos.-

El enorme esfuerzo científico y técnico desplegado para llevar a la luna a los primeros hombres merece la gratitud de la humanidad porque de ese esfuerzo se derivarán nuevos logros para aumentar el bienestar de las familias humanas. Con fe esperamos mejores días para toda la humanidad porque si al esfuerzo ya realizado se agregan luego nuevos empeños en pro de la justicia y de la libertad, como corresponden al respeto que se debe a cada ser humano y en pro de una difusión mayor del amor al prójimo, los cuales empeños podemos esperar que se hallen estimulados por el espíritu de humildad que habrá de derivarse de la conciencia más clara y vivida sobre la pequeñez de este planeta que nos sirve de hogar en los espacios siderales.-

Como representante de la Nación Costarricense hago llegar nuestro saludo a los héroes del Apolo 11 y a todos quienes han hecho posible que se realice su histórica hazaña.-

Casa Presidencial, San José, a los treinta días del mes de junio de mil novecientos sesenta y nueve.-

J. J. Trejos Fernández

I join in the wish of all Costa Ricans for the success of the historical exploit to be carried out by Apollo 11, in that it represents the scientific and technical progress attained by man in his peaceful struggle for the conquest of space and in that the crew of this ship represents human valor, will, spirit of adventure and ingenuity.

The enormous scientific and technical effort deployed in order to take the first men to the moon deserves the gratitude of mankind because from this effort will come new benefits for improving the well-being of the human race.

With faith we hope for better days for all mankind if there is later added to this successful endeavor — new determination for justice and liberty, as they correspond to the respect owed each human being and in favor of a major diffusion of love of one's neighbor, whose efforts we can hope will be stimulated by the spirit of humility derived from a more clear and vivid awareness of the minuteness of this planet, which serves as our home in the cosmos.

As representative of the Costa Rican Nation, I extend my greetings to the heroes of Apollo 11 and to all those who are making this historical feat possible.

J.J. Trejos Fernandez
President

# Cyprus

EMBASSY OF CYPRUS
WASHINGTON

Ref: P3/67　　　　　　　　　　　　　　　　July 2, 1969

Mr. T.O. Paine, Administrator,
National Aeronautics & Space
　　Administration,
400 Maryland Ave., S.W.
Room 7077,
Washington, D.C. 20546

Dear Mr. Paine,

　　Referring to your letter of June 23, 1969, addressed to the Ambassador, regarding the messages of goodwill for the Apollo 11 mission from Chief of States, I have the pleasure of quoting the message from Archbishop Makarios, President of the Republic of Cyprus.

　　" Man has conquered the moon and widened his horizons. The spaceship Apollo-11 is touching down on the surface of the moon and the first human beings are setting foot on it. The landing on the moon is the culminating achievement of a great scientific effort. That which could be captured only by the boldest imagination has now become a reality. With this historic event a new era in the life of mankind begins, and further achievements in the space world are certain to follow. We express our admiration to the valiant astronauts of Apollo-11 and to all those whose work made the conquest of space possible.

　　We join in the rejoicing of the American nation who must feel very proud because their sons were the first humans to land on the moon.

　　　　　　　　Archbishop Makarios,
　　　　　　　　President of the
　　　　　　　　Republic of Cyprus.　"

　　I regret the delay in answering your letter. I take this opportunity to express sincere wishes that the Apollo 11 mission will be a success.

　　　　　　　　　　　　　　　Yours sincerely,

Man has conquered the moon and widened his horizons. The spaceship Apollo 11 is touching down on the surface of the moon and the first human beings are setting foot on it. The landing on the moon is the culminating achievement of a great scientific effort. That which could be captured only by the boldest imagination has now become a reality. With this historic event a new era in the life of mankind begins, and further achievements in the space world are certain to follow. We express our admiration to the valiant astronauts of Apollo 11 and to all those whose work made the conquest of space possible.

We join in the rejoicing of the American nation who must feel very proud because their sons were the first humans to land on the moon.

Archbishop Makarios
President

# Dahomey

### Ambassade du Dahomey
### Washington D.C.

Le génie et l'audace d'une grande nation ouvrent aujourd'hui à l'humanité les secrets d'une planète que, depuis des siècles, elle n'a pu qu'ausculter et admirer de très loin, par le trou étroit de ses lunettes astronomiques.

Le rêve est aujourd'hui devenu réalité. Il l'est grâce au talent et aux sacrifices de la grande nation américaine. Mais il l'est aussi grâce aux produits accumulés au cours des âges et des siècles par le savoir, la science et la technique des hommes, de tous les hommes. L'épopée lunaire des astronautes d'Apollo 11 inaugure donc dans l'histoire de l'humanité un nouveau cycle grandiose d'exploits spatiaux et soulève de grandes espérances.

En ce jour de triomphe et en cette heure historique, je forme, en mon nom personnel, et en celui du Peuple dahoméen, des voeux de Paix, de Fraternité et de Bonheur pour toute l'humanité et pour tous les hommes. Que les premiers pas d'un homme sur la lune puissent convaincre les terriens de faire un jour prochain le serment de ne mettre la science et les techniques qu'au seul service de la Paix et du Progrès.

Docteur Emile-Derlin ZINSOU
Président de la République du Dahomey

The genius and daring of a great nation today open to mankind the secrets of a planet, which, for centuries, it has been able only to probe and to admire from a great distance, through the narrow hole of its telescopes.

Today the dream has been realized. It is thanks to the talent and sacrifices of the great American nation. But it is also thanks to that which has been achieved through the ages and centuries by the knowledge, science, and technology of men, of all men. The lunar epic of the Apollo 11 astronauts thus opens in the history of mankind a new, grandiose cycle of space exploits and gives rise to great hopes.

On this day of triumph and in this historic hour, I express personally, and in the name of the People of Dahomey, my wishes for Peace, Brotherhood, and Good Fortune for all mankind and for all men. May man's first steps on the moon convince those on earth to vow one day soon to employ science and technology only in the service of Peace and Progress.

Doctor Emile-Derlin Zinsou<br>
**President**

# Denmark

Denmark conveys, through the space pioneers, her warmest wishes that this spectacular landing on the moon may herald for all mankind a new era of peace and good will.

Frederik R
King of Denmark

Denmark conveys, through the space pioneers, her warmest wishes that this spectacular landing on the moon may herald for all mankind a new era of peace and good will.

Frederik R
King of Denmark

# Dominican Republic

VIA RCA
XZC YW1489 DRN0384 PTL0896 EWYWEG AG
URWA CO DRSI 145
SANTODOMINGODR 145 25 214P VIA RCA ICO
T-O- PAINE ADMINISTRADOR
NATIONAL AERONAUTIC AND SPACE ADMINISTRATION
WASHINGTONDC20546

URGENTE
4330.- EL PUEBLO DOMINICANO SIGUE CON CRECIENTE INTERES EL DESARROLLO DEL PROGRAMA DE INVESTIGACIONES ESPACIALES QUE REALIZA CON BRILLANTES RESULTADOS LA NATIONAL AERONAUTIC AND SPACE ADMINISTRATION /NASA/ COMA EL CUAL HA AMPLIADO LA VISION UNIVERSAL DE LA HUMANIDAD CONTEMPORANEA

P2/50
PUNTO PROXIMO A INICIARSE EL HISTORICO VUELO DE LA NAVE ESPACIAL APOLO 11 COMA FIJADO PARA EL 16 DE JULIO COMA CON LA MISION ESPECTACULAR DE QUE ALUNICEN DOS COSMONAUTAS COMA ME PLACE ENVIAR A LA NASA COMA MIS VOTOS FERVIENTES JUNTO CON LOS DEL PUEBLO DOMINICANO COMA PARA QUE

P/3/45
ESTA NUEVA HAZANA CIENTIFICA ALCANCE UN EXITO COMPLETO PUNTO. LA CIENCIA ESPACIAL DE LOS ESTADOS UNIDOS COMA AL SERVICIO DEL PROGRESO Y DE LA PAZ COMA OBTENDRAN UN NUEVO GALARDON EN LA EXPLORACION DE LOS ESPACIOS SIDERALES PUNTO
    JOAQUIN BALAGUER
    PRESIDENTE DE LA REPUBLICA DOMINICANA

The Dominican People follow with growing interest the development of the space exploration program being carried out by the National Aeronautics and Space Administration, which has extended the universal view of contemporary mankind.

In view of the historic flight of Apollo 11, set for July 16, with the spectacular mission of landing two men on the moon, I, together with the Dominican People, am pleased to send my best wishes to NASA that this new scientific exploit will attain complete success. Space science of the United States will reach new heights in the exploration of outer space.

Joaquin Balaguer
President

# Ecuador

```
                                       Quito, 25 de Junio de 1969


Excelentísimo Señor

Richard Nixon,

Presidente de los Estados Unidos de América:

        Saludo a Vuestra Excelencia.

        Formulo sinceros votos por el éxito de los

heroicos jóvenes astronautas que con sublime valor

pondrán sus plantas en la Luna dominando las leyes

del espacio infinito y consagrando la sublimidad de

la mente y voluntad humanas.

        Con toda consideración,

                            José María Velasco Ibarra
                            Presidente Constitucional
                            de la República del Ecuador
```

I express my sincere wishes for the success of the heroic young astronauts who, with sublime valor, will set foot on the Moon, dominating the laws of outer space and consecrating the grandeur of human understanding and goodwill.  Best regards.

Jose Maria Velasco Ibarra
**President**

# Estonia

REPUBLIC OF ESTONIA

The people of Estonia join those who hope and work for freedom and a better world.

*Ernst Jaakson*
Ernst Jaakson
Consul General of Estonia
in charge of Legation

New York, N.Y.
June 24, 1969.

The people of Estonia join those who hope and work for freedom and a better world.

Ernst Jaakson
Consul General

# Ethiopia

**Department of State TELEGRAM**

Addis Ababa Ethiopia

UNCLASSIFIED 672

PAGE 01 ADDIS 02888 301639Z

53
ACTION SCI 05

INFO OCT 01,AF 12,NASA 04,SS 20,NSC 10,P 04,USIE 00,SSO 00,NSCE 00,
INR 07,RSR 01,RSC 01,CCO 00,/065 W
--------------------  057426
O 301500Z JUN 69
FM AMEMBASSY ADDIS ABABA
TO SECSTATE WASHDC IMMEDIATE 18

UNCLAS ADDIS ABABA 2888

REF: STATE 107360

1. IEG HAS REQUESTED FOLLOWING MESSAGE BE PASSED NASA FOR USE
IN APOLLO 11 MISSION IN ACCORDANCE SUGGESTION REFTEL. IEG ALSO
CABLING TEXT SEPARATELY TO AMB MINASSIE VIA OWN CHANNELS BUT
REQUESTED OUR ASSISTANCE MAKE DEADLINE.

BEGIN TEXT TODAY'S SUCCESSFUL LUNAR LANDING IS A
MOMENTOURS OCCASION FOR ALL MANKIND. THIS MARVELOUS FEAT IS A
PROOF OF THE GIGANTIC STRIDES MAN HAS MADE IN THE FIELD OF
SCIENCE AND TECHNOLOGY. WE ARE FULLY CONFIDENT THAT THIS GREAT
MILESTONE IN MAN'S SEARCH FOR THE UNKNOWN WILL GIVE THE AMERICAN
GENIUS AND THE VALIANT AMERICAN ASTRONAUTS GREATER ENCOURAGEMENT
IN THEIR FURTHER PROBE OF THE SOLAR SYSTEM. WE ARE HOPEFUL THAT
THE RESULTS ACHIEVED IN THIS REGARD WILL ONLY BAAUSED FOR THE
WELFARE AND WELL- BEING OF MANKIND AND THE GREAT CAUSE OF WORLD
PEACE. SIGNED HAILE SELASSIE I, EMPEROR OF ETHIOPIA. END TEXT

2. EMBASSY ALSO WONDERS WHETHER FULL FORM OF MESSAGE MIGHT BE
USED IN PRESENTATION CEREMONY BETWEEN HIM AND ASTRONAUTS AT CAPE
KENNEDY AT TIME HIM VISIT NEXT MONTH. IDEA MENTIONED INFORMALLY
BUT NOT CLEARED WITH IEG. APPRECIATE DEPARTMENT'S COMMENTS.

YOST

UNCLASSIFIED

Today's successful lunar landing is a momentous occasion for all mankind. This marvelous feat is a proof of the gigantic strides man has made in the field of science and technology. We are fully confident that this great milestone in man's search for the unknown will give the American genius and the valiant American astronauts greater encouragement in their further probe of the solar system. We are hopeful that the results achieved in this regard will only be used for the welfare and well-being of mankind and the great cause of the world peace.

**Haile Selassie I**
**Emperor**

# Ghana

CHAIRMAN
NATIONAL LIBERATION COUNCIL

THE CASTLE
OSU, ACCRA

24th June, 1969.

MESSAGE TO BE TAKEN TO THE MOON
BY THE U.S. ASTRONAUTS FOR THE
FIRST MOON LANDING - 1969

WE PRAY THAT YOUR HISTORIC LANDING ON THE MOON MAY USHER IN AN ERA OF PEACE AND PROSPERITY AND GOODWILL AMONG ALL MEN HERE ON EARTH.

(Brigadier A.A. AFRIFA, D.S.O.)
CHAIRMAN, N.L.C. OF GHANA.

We pray that your historic landing on the moon may usher in an era of peace and prosperity and goodwill among all men here on earth.

Brigadier A.A. Afrifa, D.S.O.
Chairman, National Liberation Council

# Great Britain

Message from Her Majesty Queen Elizabeth II

On behalf of the British people I salute the skill and courage which have brought man to the moon. May this endeavour increase the knowledge and well-being of mankind.

Elizabeth R.

**Message from Her Majesty Queen Elizabeth II**

On behalf of the British people I salute the skill and courage which have brought man to the moon.  May this endeavor increase the knowledge and well-being of mankind.

                                                                       **Elizabeth R.**

# Greece

ROYAL GREEK EMBASSY
WASHINGTON, D.C.

MESSAGE
OF GOOD WILL FOR THE APOLLO 11 ASTRONAUTS
BY H. E. THE REGENT OF GREECE

At this historical moment, when man lands on the moon, I express on behalf of the Greek people my heartfelt congratulations to the Government of the U.S.A. and the distinguished men of the American nation, who led the way towards new horizons for the human race.

I am particularly proud speaking on behalf of the Greek nation, whose ancestors had the privilege to be forerunners in the philosophical thought and scientific research, which first penetrated the universe.

It is a happy coincidence that the amazing program of man's flight to space, which has been so magnificently fulfilled today, bears the name of the Greek God Apollo; this symbolic name demonstrates the never ending effort of man to achieve knowledge, beyond time and place.

At this historical moment, when man lands on the moon, I express on behalf of the Greek people my heartfelt congratulations to the Government of the U.S.A. and the distinguished men of the American nation, who led the way towards new horizons for the human race.

I am particularly proud speaking on behalf of the Greek nation, whose ancestors had the privilege to be forerunners in the philosophical thought and scientific research, which first penetrated the universe.

It is a happy coincidence that the amazing program of man's flight to space, which has been so magnificently fulfilled today, bears the name of the Greek God Apollo; this symbolic name demonstrates the never ending effort of man to achieve knowledge, beyond time and place.

The difficulties, which had once broken the wings of Ikarus, are surpassed by man's persistence in his search for truth, and he is staring from the moon at the earth, whose peace and welfare should be now, more than ever, his main preoccupation.

George Zoitakis
Lieutenant General Regent

# Guyana

Georgetown
Guyana,
South America.

## TEXT OF MESSAGE RECEIVED FROM THE PRIME MINISTER OF GUYANA ON THE OCCASION OF MAN'S FIRST LANDING ON THE MOON

### - To Those Coming After -

We cannot tell on what future day - beings of our own kind or perhaps from some other corner of the cosmos, will come upon this message but for those coming after, we wish to record three things:

First, we salute these astronauts, the first two of our human race who with faith and courage have voyaged far beyond the familiar limits of our earthly home to the moon. It is certain that their mission ushers in the greatest adventure of life since its primaeval beginnings on this planet, Earth.

Second, as members of our human race thus thrust among the stars, we pledge ourselves to work towards ensuring that the technology which has made it possible and the resources which may be discovered will be used for the benefit of all mankind irrespective of terrestrial divisions of race or creed or levels of development.

Third and finally, we wish to set down the facts about the people for whom I speak. We are a small nation of some 700,000 souls living on the shoulder of South America in a country some 83,000 square miles in area. Our ancestors came from nearly every corner of the planet Earth and our people today profess a variety of creeds and of ways of living. But in a world in which divisions deepen and where too often one man's hand is set against his brother, we are proud

/that ...

2.

that we have given to our time an example of how out of diversity we have made one people, one nation - with one destiny.

In working out this destiny, we have developed institutions based on the recognition of the equality of all men, forms of government in which all can participate and a system of justice which protects the weak. With the help of friendly nations, and working together, we are embarked on the challenging task of abolishing disease and poverty from our midst, and of developing our economy so that it can support a worthy level of living for our people. We have, likewise, striven hard to ensure that men everywhere are free to determine their own way of life.

We do not know what shall be the judgment of history but we would be well pleased if on some later day when this is read, it is said of us that we strove greatly to advance the dignity of all men.

Linden Forbes Sampson Burnham
Prime Minister of Guyana

To those coming after: We cannot tell on what future day — beings of our own kind or perhaps from some other corner of the cosmos, will come upon this message but for those coming after, we wish to record three things:

First, we salute these astronauts, the first two of our human race who with faith and courage have voyaged far beyond the familiar limits of our earthly home to the Moon. It is certain that their mission ushers in the greatest adventure of life since its primeval beginnings on this planet, Earth.

Second, as members of our human race thus thrust among the stars, we pledge ourselves to work towards ensuring that the technology which has made it possible and the resources which may be discovered will be used for the benefit of all mankind irrespective of terrestrial divisions of race or creed or levels of development.

Third and finally, we wish to set down the facts about the people for whom I speak. We are a small nation of some 700,000 souls living on the shoulder of South America in a country some 83,000 square miles in area. Our ancestors came from nearly every corner of the planet Earth and our people today profess a variety of creeds and of ways of living. But in a world in which divisions deepen and where too often one man's hand is set against his brother, we are proud that we have given to our time an example of how out of diversity we have made one people, one nation — with one destiny.

In working out this destiny, we have developed institutions based on the recognition of the equality of all men, forms of government in which all can participate and a system of justice which protects the weak. With the help of friendly nations, and working together, we are embarking on the challenging task of abolishing disease and poverty from our midst, and of developing our economy so that it can support a worthy level of living for our people. We have, likewise, striven hard to ensure that men everywhere are free to determine their own way of life.

We do not know what shall be the judgment of history but we would be well pleased if on some later day when this is read, it is said of us that we strove greatly to advance the dignity of all men.

Linden Forbes Sampson Burnham
Prime Minister

# Iceland

Reykjavík, 30. júní 1969.

Íslenzka þjóðin sendir kveðju sína með
Appollo 11. og óskar geimförunum velfarnaðar á
sögulegri ferð þeirra. Megi afrek geimvísindanna
boða tíma friðar og hamingju öllu mannkyni.

Kristján Eldjárn
forseti Íslands

The people of Iceland convey their greetings by Apollo 11 and wish the astronauts good fortune on their historic voyage. May the great achievements of space research inaugurate an era of peace and happiness for all mankind.

**Kristjan Eldjarn**
**President**

# India

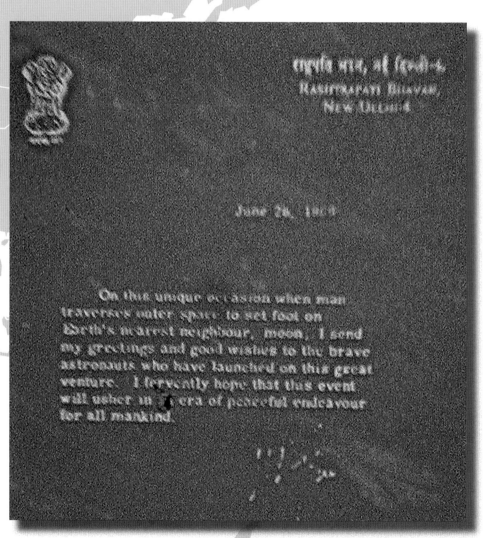

*Image distortion due to microscope limitations.*

On this unique occasion when man traverses outer space to set foot on Earth's nearest neighbor, Moon, I send my greetings and good wishes to the brave astronauts who have launched on this great venture. I fervently hope that this event will usher in an era of peaceful endeavor for all mankind.

Indira Gandhi
Prime Minister

# Iran

On this occasion when Mr. Neil Armstrong and Colonel Edwin Aldrin set foot for the first time on the surface of the Moon from the Earth, we pray the Almighty God to guide mankind towards ever increasing success in the establishment of peace and the progress of culture, knowledge and human civilization.

**Mohammad Reza Pahlavi Aryamehr**
**Shahanshah (the Shah of Iran)**

# Ireland

**Uachtarán na hÉireann**
PRESIDENT OF IRELAND

BAILE ÁTHA CLIATH 8
DUBLIN

Go ndeonaí Dia go dtabharfaidh an mheabhair agus an misneach a chuir ar chumas an duine cos a leagan ar an ngealach go mbeidh ar a chumas chomh maith síocháin agus sonas a chur in áirithe ar an talamh seo agus teacht slán ó chontúirt a léirscriosta féin.

*Eamon de Valera*

May God grant that the skill and courage which have enabled man to alight upon the Moon will enable him, also, to secure peace and happiness upon the Earth and avoid the danger of self-destruction.

Eamon de Valera
President

# Israel

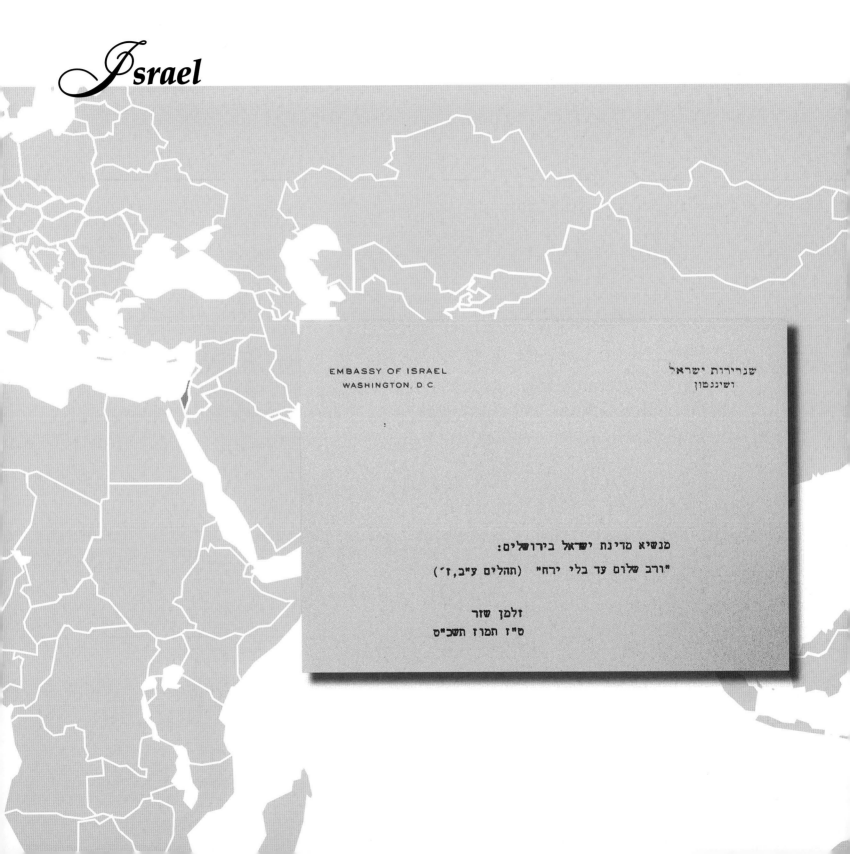

EMBASSY OF ISRAEL
WASHINGTON, D.C.

שגרירות ישראל
וושינגטון

מנשיא מדינת ישראל בירושלים:

"וירב שלום עד בלי ירח" (תהלים ע"ב, ז')

זלמן שזר

ס"ז תמוז תשכ"ט

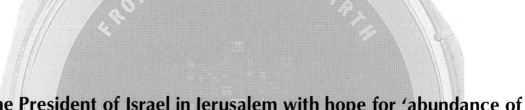

**From the President of Israel in Jerusalem with hope for 'abundance of peace so long as the Moon endureth'** (Psalms 72,7).

                                        Zalman Shazar

# Italy

Il Presidente
della Repubblica Italiana

[illegible]

La conquista della Luna segna una tappa gloriosa nel cammino di tutta l'umanità verso la pace, la libertà, la giustizia.

Giuseppe Saragat

[illegible] luglio 1969

The courage and the technology of the United States of America have brought to our satellite this message of the Head of the Italian Nation which prides itself to number amongst its sons Galileo Galilei, whose genius paved the ways for modern science.

The conquest of the Moon is a glorious milestone along the road of all mankind towards the achievement of peace, freedom and justice.

Guiseppe Sarget
President

# Ivory Coast

ABIDJAN, 1er JUILLET 1969

*Président de la République de Côte d'Ivoire*

AU MOMENT OÙ LE PLUS VIEUX REVE DES HOMMES DEVIENT UNE REALITE, JE SUIS TRES SENSIBLE A LA DELICATE ATTENTION QU'A EUE LA NASA EN ME PROPOSANT LES SERVICES DU PREMIER MESSAGER HUMAIN QUI VA TOUCHER LA LUNE POUR Y PORTER LA PAROLE DE LA COTE D'IVOIRE.

JE VOUDRAIS QUE CE PASSAGER DU CIEL, LORSQU'IL MARQUERA LE SOL LUNAIRE DE L'EMPREINTE DE L'HOMME, RESSENTE À QUEL POINT NOUS SOMMES FIERS D'APPARTENIR À LA GENERATION QUI A ACCOMPLI CET EXPLOIT.

JE VOUDRAIS AUSSI QU'IL DISE À LA LUNE COMBIEN ELLE EST BELLE QUAND ELLE ILLUMINE LES NUITS DE LA COTE D'IVOIRE.

JE VOUDRAIS SURTOUT QU'IL TOURNE LA TETE VERS NOTRE PLANETE TERRE ET QU'IL LUI CRIE COMBIEN LES PROBLEMES QUI DECHIRENT LES HOMMES SONT INSIGNIFIANTS QUAND ON LES VOIT DE LA-BAS.

PUISSE SA PAROLE, EN DESCENDANT DU CIEL, TROUVER DANS LE COSMOS LA FORCE ET LA LUMIERE QUI LUI PERMETTRONT DE CONVAINCRE L'HUMANITE DE LA BEAUTE DU PROGRES DANS LA FRATERNITE ET LA PAIX.

Félix HOUPHOUET-BOIGNY

At the moment when man's oldest dream is becoming a reality, I am very thankful for NASA's kind attention in offering me the services of the first human messenger to set foot on the Moon and carry the words of the Ivory Coast.

I would hope that when this passenger from the sky leaves man's imprint on lunar soil, he will feel how proud we are to belong to the generation which has accomplished this feat.

I hope also that he would tell the Moon how beautiful it is when it illuminates the nights of the Ivory Coast.

I especially wish that he would turn towards our planet Earth and cry out how insignificant the problems which torture men are, when viewed from up there.

May his work, descending from the sky, find in the Cosmos the force and light which will permit him to convince humanity of the beauty of progress in brotherhood and peace.

Felix Houphouet-Boigny
President

## Jamaica

EMBASSY OF JAMAICA
1666 CONNECTICUT AVENUE, N.W.
WASHINGTON, D. C., 20009

TELEPHONE 232-1036

MAY HE WHOSE GLORY THE HEAVENS DECLARE

GRANT THAT MANKIND MAY GROW IN THE

KNOWLEDGE OF HIS PURPOSES AS WE PROBE

INTO THE SECRETS OF HIS UNIVERSE.

(SGD) HUGH LAWSON SHEARER
PRIME MINISTER OF
JAMAICA.

May He whose glory the heavens declare grant that mankind may grow in the knowledge of His purposes as we probe into the secrets of His universe.

**Hugh Lawson Shearer**
**Prime Minister**

# Japan

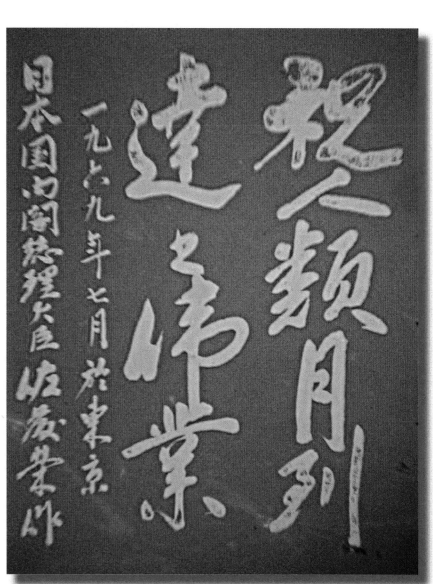

**In congratulations of the outstanding achievement of humanity's arrival on the Moon.**

Eisaku Sato
Prime Minister

# Kenya

Department of State

**TELEGRAM**

UNCLASSIFIED 387

PAGE 01 NAIROB 03073 021538Z

52
ACTION SCI 05

INFO OCT 01, AF 12, NASA 04, SS 20, NSC 10, P 04, USIA 12, IO 13, INR 07,
    SSO 00, /088 W
                                                    004735

P 021402Z JUL 69
FM AMEMBASSY NAIROBI
TO SECSTATE WASHDC PRIORITY 8679

UNCLAS NAIROBI 3073

SUBJECT: GOK MESSAGE FOR APOLLO-11

REFERENCE: STATE 107360

1. IN LETTER DATED JULY 2 MFA HAS SENT FOLLOWING
MESSAGE (IN SWAHILI) FROM PRESIDENT KENYATTA TO
BE CARRIED ON APOLLO 11 MISSION AND DEPOSITED
ON LUNAR SURFACE:

QUOTE KWA MINTARAFU YA WATU WOTE WA KENYA,
NAWAPA TAHNIA WATU WA AMERICA KWA KUWEZA KUFIKA
MWEZINI. NI JAMBO LA KUTIA MOYO SANA KUWA SISI
BINAADAMU TUMEWEZA KUFIKA MWEZINI. NI DHAHIRI KUWA
SISI SISI SOTE NI NDUGU HAPA ULIMWENGUNI
NA NI WAJIBU WETU KUSHIRIKIANA KWA MAMBO YOTE.

2. OUR INFORMAL TRANSLATION FOLLOWS (MESSAGE
SHOULD, OF COURSE, BE IN SWAHILI):
QUOTE ON BEHALF OF ALL THE PEOPLE OF KENYA, I
CONGRATULATE THE PEOPLE OF AMERICA FOR ACCOMPLISHING ARRIVAL
ON THE MOON. IT IS A VERY INSPIRING EVENT FOR ALL
MANKIND THAT WE HAVE BEEN ABLE TO REACH THE MOON.
IT IS CLEAR THAT WE ALL ARE BROTHERS HERE ON EARTH AND
THAT IS OUR OBLIGATION TO COOPERATE TOGETHER IN ALL
ENDEAVORS. UNQUOTE

3. GOK APOLOGIZES FOR LATENESS OF MESSAGE, BUT HOPES,
AS DOES EMBASSY, THAT IT CAN BE INCLUDED. MESSAGE
UNQUESTIONABLY HAS FULL KNOWLEDGE AND APPROVAL OF

On behalf of all the people of Kenya, I congratulate the people of America for accomplishing arrival on the Moon. It is a very inspiring event for all mankind that we have been able to reach the Moon. It is clear that we all are brothers here on Earth and that it is our obligation to cooperate together in all endeavors.

President Kenyatta

# Korea

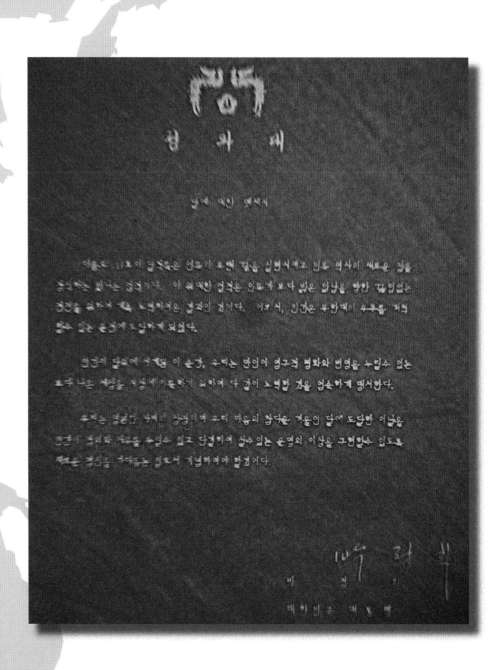

The landing on the Moon by Apollo 11 is a brilliant feat of all mankind which makes men's dreams a reality and marks a new chapter of human history. This great achievement is a result of man's constant striving for progress towards a brighter destiny. Now, realization of man's adventure into yet further reaches of space seems but a few steps away.

On this historic occasion, we do solemnly pledge ourselves to work together on this Earth for the better world with lasting peace and prosperity for all mankind. Let us celebrate the first landing of men on the Moon, the symbol of eternal grace and the mirror of man's true heart, with a new spirit which will inspire mankind to realize the ideal of civilization in which men live in justice, freedom, and unity.

**Park Chung Hee**
**President**

# *L*aos

EN TANT QUE REPRÉSENTANT D'UN PEUPLE ET D'UNE NATION DE LA TERRE JE M'ASSOCIE A LA FIERTÉ INCOMMENSURABLE ÉPROUVÉE ET JE PARTAGE L'ÉMOTION INTENSÉMENT RESSENTIE PAR LE PEUPLE AMÉRICAIN ET LA NATION AMÉRICAINE DANS LE PREMIER CONTACT HUMAIN AVEC LA PLANÈTE LUNE.

SRI SAVANG VATTHANA
KING OF LAOS

As the representative of a people and a nation of the Earth, I feel the immeasurable pride and I share the intensely felt emotion of the American people and the American Nation in the first human contact with the lunar planet.

SRI (HM) Savang Vatthana

# Latvia

LATVIAN LEGATION
WASHINGTON, D C

July 1, 1969.

Dear Mr. Paine:

    With reference to our previous correspondence I wish to offer the following message to be taken to the moon by Apollo astronauts:

    ON BEHALF OF THE LATVIAN NATION I SALUTE THE FIRST MEN ON THE MOON AND PRAY FOR THEIR SAFE RETURN.

    MAY THEIR ACHIEVEMENT CONTRIBUTE TO WORLD PEACE AND RESTORATION OF FREEDOM TO ALL NATIONS.

A. Dinbergs
Counselor

Mr. T.O. Paine, Administrator
National Aeronautics and
   Space Administration
      Washington, D.C. 20546

On behalf of the Latvian nation I salute the first men on the Moon and pray for their safe return.

May their achievement contribute to world peace and restoration of freedom to all nations.

A. Dinbergs
Counselor

# Lebanon

**EMBASSY OF LEBANON**
**WASHINGTON**

Message de Son Excellence Monsieur Charles Helou, President de la Republique Libanaise, pour etre depose sur la lune par les Astronautes D'Apollo II.

--------

Au terme d'une conjonction d'Energies, de Sciences et de Techniques, toutes a l'honneur de ses promoteurs, l'Homme atteint la lune et y fait ses premiers pas. A ce moment glorieux de liberation des contraintes physiques de la condition humaine, le Liban joint son temoignage vecu de Terre de Rencontre et de Coexistence de Familles Spirituelles Multiples pour redire sa foi dans la promotion de l'Homme a travers les valeurs de Paix, de Justice et de Liberte.


Baabda, le 28 Juin 1969

CHARLES HELOU

As a final result of the combined efforts of science and technology and all to the honor of its initiators, man is reaching the Moon and is setting foot on it for the first time. At this glorious moment when man is freed from the physical constraints of the human condition, Lebanon adds her living testimonial as a land of encounter and the coexistence of many spiritual families to reaffirm her faith in the advancement of man through the virtues of peace, justice and of liberty.

Charles Helou
President

# Lesotho

**EMBASSY OF THE KINGDOM OF LESOTHO**

1716 NEW HAMPSHIRE AVE., N.W.
WASHINGTON, D.C. 20009
TELEPHONE 462-4190

July 1, 1969

Dear Mr. Paine,

I have been instructed to forward the following message from the Honourable Prime Minister Chief Leabua Jonathan to the Government of the United States:

"THE GOVERNMENT AND PEOPLE OF LESOTHO WISH THE GOVERNMENT AND PEOPLE OF THE UNITED STATES OF AMERICA EVERY SUCCESS IN THEIR ATTEMPT TO LAND THE FIRST HUMAN BEING ON THE MOON. OUR SINCERE WARM WISHES FOR A SAFE JOURNEY TO AND FROM THE MOON TO APOLLO 11 ASTRONAUTS."

Sincerely yours,

Mothusi T. Mashologu,
Ambassador

Mr. T. O. Paine,
Administrator,
N.A.S.A.,
Washington, D.C.

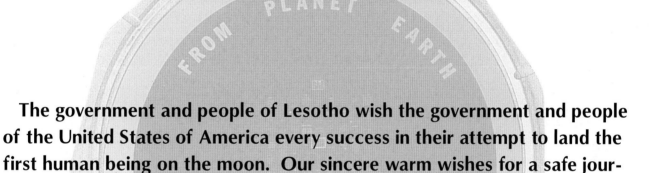

The government and people of Lesotho wish the government and people of the United States of America every success in their attempt to land the first human being on the moon. Our sincere warm wishes for a safe journey to and from the moon to Apollo 11 Astronauts.

> Leabua Jonathan
> Prime Minister

# Liberia

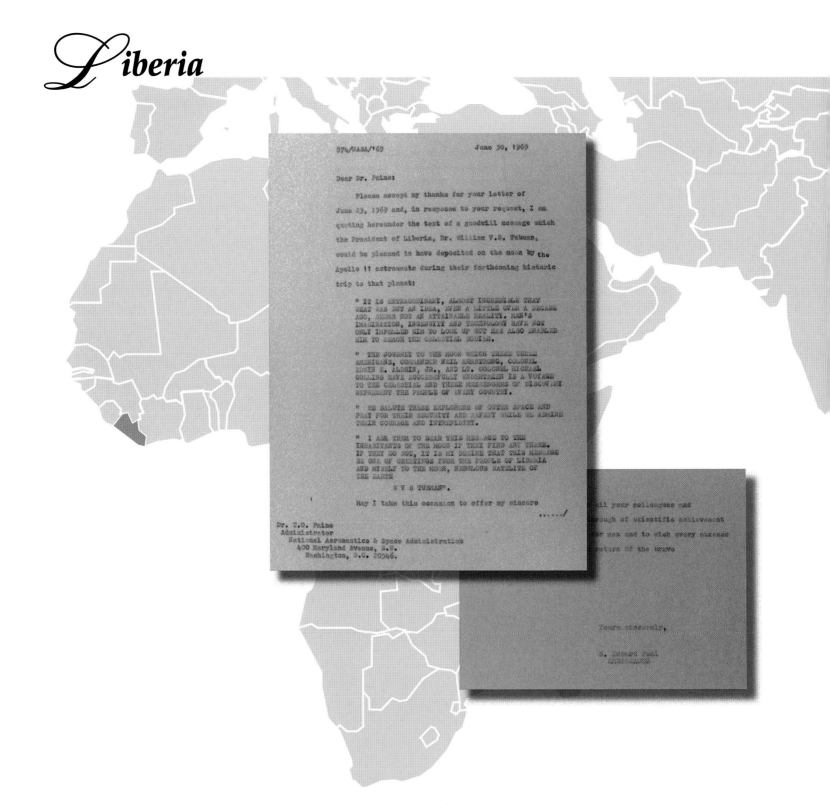

It is extraordinary, almost incredible that what was but an idea, even a little over a decade ago, seems now an attainable reality. Man's imagination, ingenuity and technology have not only impelled him to look up but has also enabled him to reach the celestial bodies.

The journey to the Moon which these three Americans, Commander Neil Armstrong, Colonel Edwin E. Aldrin, Jr., and Lt. Colonel Michael Collins have successfully undertaken is a voyage to the celestial and these messengers of discovery represent the people of every country.

We salute these explorers of outer space and pray for their security and safety while we admire their courage and intrepidity.

I ask them to bear this message to the inhabitants of the Moon if they find any there. If they do not, it is my desire that this message be one of the greetings from the people of Liberia and myself to the Moon, Nebulous satellite of the Earth.

W.V.S. Tubman<br>
President

**REPOBLIKA MALAGASY**
Fahafahana - Tanindrazana - Fandrosoana

WASHINGTON, D. C.

AMIN'IZAO TAON-JATO FAHA-ROAPOLO KA NAHATONGAVAN'NY ZANAK'OLOMBELONA VOALOHANY ENY AMIN'NY VOLONA, DIA MANIRY INDRINDRA ANDRIAMATOA PHILIBERT TSIRANANA, FILOHA NY REPOBLIKA MALAGASY SY NY VAHOAKA TSY VAKY VOLO MANERANA AN'I MADAGASIKARA MBA HITONDRA FAHAFAHANA, FIADANANA ARY FANDROSOANA HOAN'IZAO TONTOLO IZAO NY ZAVA-BITA MAHATALANJONA ATERAKY NY FAHAIZANA SY NY HERIM-PO ETO AMBONIN'NY TANY.

ENGA ANIE IZANY ZAVA-BITA IZANY SY NY FITARANY ARY NY VOKANY MBA TSY HO FIASANA HANIMBA NA HITERAKA FAHALEVONANA FA HITONDRA KOSA FAHASOAVANA ARA-BATANA, ARA-TSAINA, ARA-PANAHY HO AN'NY MIAINA REHETRA.

FANIRINA VELONA AO AM-PON'NY MALAGASY TSY VAKY VOLO IZANY KA IRINY HITOETRA MAHARITRA ENY AMIN'NY VOLANA ARY AMPITONDRAINY IREO MPITONDRA SAMBONDANITRA AMERIKANA APOLLO FAHA-XI.

HO ELA VELONA ANIE IZAO TONTOLO IZAO MIANKINA AMIN'NY FAHENDRENA SY FAHAIZANA NOMEN'ANDRIAMANITRA NY ZANAK'OLOMBELONA.

NATAO TETO ANTANANARIVO TAMIN'NY 30 JIONA 1969

AVY AMIN'NY VAHOAKAN'I MADAGASIKARA SY NY FILOHANY

PHILIBERT TSIRANANA

In this twentieth century when for the first time man has reached the Moon, Mr. Philibert Tsiranana, President of the Malagasy Republic, and the people of Madagascar sincerely wish that the marvelous accomplishments of the knowledge and courage of man will bring the world liberty, peace and progress.

May these accomplishments, their development and their consequences not be transformed into instruments of destruction but may they bring physical, intellectual and moral well-being to all living beings.

Such is the sincere desire of the Malagache people that they would like transmitted to the Moon by the American astronauts of Apollo 11.

Long live the world founded on the wisdom and knowledge that God has given to man.

Philibert Tsiranana
President

# Malaysia

Message from His Majesty The Yang di-Pertuan Agong of Malaysia :-

" The people of Malaysia join the rest of the world today in congratulating the Government and people of the United States of America on the success of the Apollo 11 mission to land man for the first time on the moon. May the knowledge gained in the efforts to fulfill this historic mission add to the wisdom of mankind in our search for greater peace and prosperity.

                           ISMAIL NASIRUDDIN SHAH
                           YANG DI-PERTUAN AGONG OF MALAYSIA."

Tan Sri Ong Yoke Lin

The people of Malaysia join the rest of the world today in congratulating the Government and people of the United States of America on the success of the Apollo 11 mission to land man for the first time on the moon. May the knowledge gained in the efforts to fulfill this historic mission add to the wisdom of mankind in our search for greater peace and prosperity.

Ismail Nasiraddin Shah

# Maldives

THIS MESSAGE OF PEACE AND GOODWILL FROM THE PEOPLE OF MALDIVES CAME WITH THE FIRST MEN FROM PLANET EARTH TO SET FOOT ON THE MOON.

IBRAHIM NASIR
PRESIDENT OF THE REPUBLIC
OF MALDIVES

**This message of peace and goodwill from the people of Maldives came with the first men from planet Earth to set foot on the Moon.**

**Ibrahim Nasir**
**President**

# Mali

AMBASSADE
DE LA REPUBLIQUE DU MALI
2130 R STREET, N.W.
WASHINGTON, D. C. 20008
TELEPHONE DECATUR 2-2249

Message sent by the Chief of State of the Republic of Mali for the Apollo 11 astronauts to deposit on the moon :

"Au nom du Peuple et du Gouvernement de la République du Mali, je tiens à rendre un chaleureux hommage à tous ceux qui par leur intelligence et leur courage ont permis à l'homme d'atterrir sur la lune, ouvrant ainsi à l'humanité un nouvel horizon plein de promesses.

Notre Peuple et notre Gouvernement souhaitent que cette étape historique dans la marche vers le progrès contribue essentiellement au renforcement de la paix, au rapprochement de tous les hommes et à la liquidation de la misère sur notre planète.

Lieutenant Moussa TRAORE
Président du Comité Militaire
de Libération Nationale
de la République du Mali,
Chef de l'Etat

Bamako, le 30 juin 1969 "

In the name of the People and the Government of the Republic of Mali, I wish to express my warmest respect to all those who by their intelligence and by their courage have permitted man to land on the Moon, thus opening to humanity a new horizon full of promise.

Our People and our Government hope that this historic step in the march towards progress will contribute essentially to the reinforcement of peace, to the bringing together of all men and to the liquidation of misery on our planet.

Lieutenant Moussa TRAORE
Chief of State

# Malta

**EMBASSY OF MALTA**
WASHINGTON, D.C. 20008

ON THIS UNIQUE AND HISTORIC OCCASION WHEN MAN FIRST SET FOOT ON A PLANET OUTSIDE HIS OWN, THE PEOPLE OF MALTA JOIN THE REST OF THE WORLD IN SALUTING THE MEN WHOSE COURAGE AND DEDICATION, BACKED BY THE UNTIRING EFFORTS OF SCIENTISTS AND COUNTLESS COLLABORATORS, HAVE MADE POSSIBLE THIS NEW CONQUEST IN SPACE AND IN THE SAME WAY AS MALTA HAS ADVOCATED PEACE BELOW THE WATERS OF THE WORLD SHE FERVENTLY PRAYS THAT PEACE SHALL CONTINUE TO REIGN IN THE VASTNESS OF SPACE BEYOND IT.

GIORGIO BORG OLIVIER
PRIME MINISTER

On this unique and historic occasion when man first set foot on a planet outside his own, the people of Malta join the rest of the world in saluting the men whose courage and dedication, backed by the untiring efforts of scientists and countless collaborators, have made possible this new conquest in space and in the same was as Malta has advocated peace below the waters of the world she fervently prays that peace shall continue to reign in the vastness of space beyond it

Giorgio Borg Olivier
Prime Minister

# Mauritius

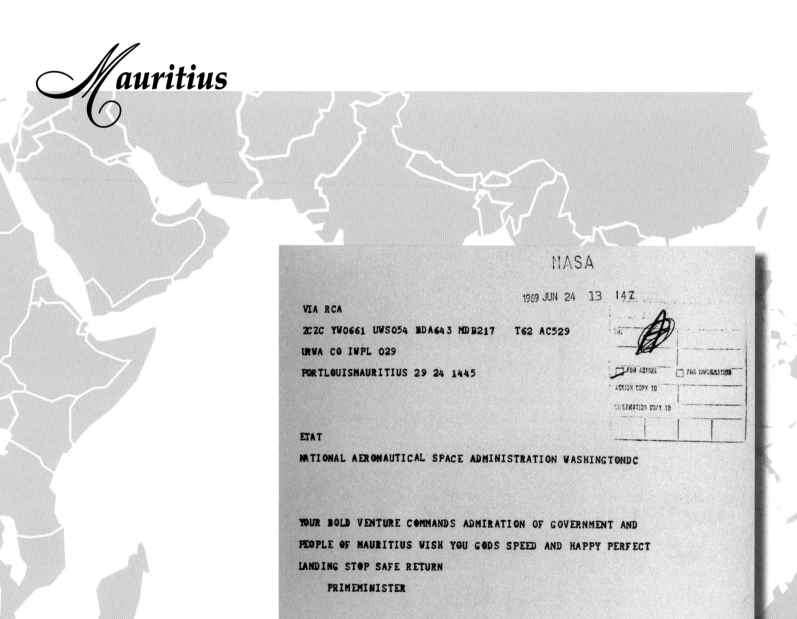

## Apollo 11 Goodwill Messages 187

**Your bold venture commands admiration of government and people of Mauritius wish you God's speed and happy perfect landing. Safe return.**

**Seewoosagar Ramgoolam
Prime Minister**

# Mexico

PRESIDENCIA DE LA REPUBLICA

Es para México un honor el participar, con este modestísimo, simbólico testimonio, en la aventura de llevar al hombre, por primera vez, a un suelo distinto de su planeta. Al hacerlo, formula su más entusiasta felicitación a los esforzados e intrépidos astronautas, a los hombres de ciencia, a los técnicos y, de una manera más amplia, al pueblo norteamericano y a sus gobernantes por esta empresa que sólo tiene antecedentes en la historia de nuestra fantasía.

El propio emblema entrañable de México —su escudo tradicional— recoge ya, en el águila y la serpiente, el doble signo que ha inspirado al hombre desde sus remotos orígenes, y que se antoja, de modo especial, el de la humanidad por venir: la serpiente, que es cifra de la tierra y de cuanto en ella nos retiene; y el águila, que figura el vuelo audaz y avizor, el arrojado peregrinaje que hace posible acrecer con ámbitos siempre nuevos el legado de los siglos. Lejos de ser contradictorias, ambas imágenes se complementan, expresando, a un tiempo,

nuestra condición carnal

— 2 —

PRESIDENCIA DE LA REPUBLICA

nuestra condición carnal y terrestre y los sueños de que se alimenta todo progreso.

En 1492, el descubrimiento del Continente Americano transformó la geografía y el rumbo del acontecer humano. Hoy, la conquista del espacio extraterrestre, —con las incógnitas que allí nos aguardan—, renueva nuestras perspectivas y agiganta nuestros paradigmas. México, al expresar su esperanza de que esta hazaña del hombre redunde en beneficio del hombre y de que en ella llegarán a participar, con limpia conciencia de la comunidad de su destino, todos los pueblos de la tierra, ofrece aportar al desenvolvimiento de la nueva etapa, no un poderío o una riqueza que no tiene, pero sí la herencia moral que ha decantado su historia: una sed insaciable de mejoramiento material y espiritual y una sólida fe en la preponderancia de la razón y la justicia como guía e inspiración de la conducta humana que asume una nueva, ingente responsabilidad.

México, D. F., 26 de junio de 1969.

GUSTAVO DIAZ ORDAZ.
Presidente de los Estados Unidos Mexicanos.

It is an honor for Mexico, with this most modest symbolic testimony, to form part of the event which for the first time takes man to a soil away from his home planet. And, it doing so, Mexico extends most enthusiastic congratulations to the dedicated, gallant astronauts and to the scientists and technicians, as well as, in a broader sense, to the American people and their Government for this undertaking that, hitherto, only had precedents in the realm of imagination.

Mexico's very own emblem-its traditional seal-with the eagle and the serpent, already embodies the double sign inspiring man since his remote origins and which in a particular manner may be equated to coming humanity: the serpent represents flight, undaunted and far-seeing, a fearless pilgrimage which makes it possible for the legacy of the centuries to reach ever increasing circling horizons. Far from being contradictory to each other, both images are complementary and placed together reflect our temporal, earthly, nature and the visions which nurture all progress.

In 1492, the discovery of the American Continent transformed geography and the course of human events. Today, conquest of ultraterrestrial space - with its attendant unknowns — recreates our perspectives and enhances our paradigms.

Mexico, while expressing its hope that this human accomplishment will result in good for mankind and that all the peoples on Earth will participate in its fulfillment with clear conscience of their common destiny, for the development of this new stage, offers not a power nor richness it does not possess but the moral heritage decanted from its own history: an unquenchable thirst for material and spiritual improvement and an unyielding faith in the supremacy of reason and justice as a way and an inspiration for human conduct which now has attained a new far reaching responsibility.

Gustavo Díaz Ordaz<br>
**President**

# Morocco

The Ambassador of Morocco

June 26, 1969

His Majesty King Hassan II and the people of Morocco wish to join the other nations of the world in saluting the courage of the first men to set foot on the moon in the spirit of peace for all mankind. May this spirit of peace pervade the earth and the advancement of science enrich the great brotherhood of men.

Ahmed Osman
Ambassador

His Majesty King Hassan II and the people of Morocco wish to join the other nations of the world in saluting the courage of the first men to set foot on the Moon in the spirit of peace for all mankind. May this spirit of peace pervade the Earth and the advancement of science enrich the great brotherhood of men.

**Ahmed Osman
Ambassador**

# The Netherlands

The Netherlands

Ik heb grote bewondering voor de vaardigheid
en de volharding van allen, die ertoe hebben bij-
gedragen om de eerste bemande vlucht naar de maan
mogelijk te maken. Ik hoop, dat deze prestatie
een zegen zal blijken te zijn voor de toekomst
van de gehele mensheid.

Juliana R.

I have great admiration for the skill and perseverance of all those who have contributed to make the first manned flight to the Moon possible. I hope that this achievement will prove of great benefit for the future of mankind.

Juliana R.

Prime Minister
Wellington
New Zealand

By this flight man has finally fulfilled the great ambition of setting foot on another celestial body. As Prime Minister of New Zealand I hope that the realisation of this dream — so long remote — will inspire all those who set their sights high and thus bring closer the dreams we share of peace and cooperation for all mankind.

Keith Holyoake

Prime Minister of New Zealand

By this flight man has finally fulfilled the great ambition of setting foot on another celestial body.  As Prime Minister of New Zealand I hope that the realization of this dream -- so long remote — will inspire all those who set their sights high and thus bring closer the dreams we share of peace and cooperation for all mankind.

              Keith J. Holyoake

# Nicaragua

PRESIDENCIA DE LA REPÚBLICA
MANAGUA, D. N. NICARAGUA, C. A.

DE GRAN TRASCENDENCIA PARA LA PAZ, PARA LA
INVESTIGACION CIENTIFICA DEL ORIGEN DE LA
TIERRA Y DEL SISTEMA PLANETARIO SOLAR, SERA
LA LLEGADA DEL HOMBRE A LA LUNA. ESTE ACTO
EXTRAORDINARIO Y ESTE TRIUNFO DEL HOMBRE
APLICANDO LA CIENCIA, NOS INSPIRA A PENSAR
EN EL CREADOR.

EL PUEBLO DE NICARAGUA FORMULA SUS MAS
FERVIENTES VOTOS POR EL EXITO DEL VUELO
ESPACIAL DEL APOLO XI Y PATENTIZA SU SINCERO
Y PROFUNDO RECONOCIMIENTO AL PUEBLO Y
GOBIERNO DE LOS ESTADOS UNIDOS DE AMERICA,
A LOS ASTRONAUTAS NEIL ARMSTRONG, EDWIN E.
ALDRIN Y MICHAEL COLLINS, QUE CON SUS ESFUERZOS
HARAN POSIBLE LA INCORPORACION DE LA LUNA.

A. SOMOZA

Managua, D.N. 27 de Junio de 1969.

The arrival of man on the moon will be of great consequence for peace and for the scientific investigation of the origin of the Earth and the Solar System.  This extraordinary event and this triumph of man in the application of science inspires us to think of the Creator.

The People of Nicaragua express their most fervent wishes for the success of the flight of Apollo 11 and show their sincere and profound recognition of the People and government of the United States of America and Astronauts Neil Armstrong, Edwin E. Aldrin, and Michael Collins, who, by their efforts will make possible the conquest of the Moon.

A. Somoza

## Norway

OSLO SLOTT
June 1969

I express my best wishes for the Astronauts carrying out the Apollo 11 mission and for the success of this historic space journey.

I express my best wishes for the Astronauts carrying out the Apollo 11 mission and for the success of this historic space journey.

Olav R.

# Pakistan

PRESIDENT'S HOUSE,
RAWALPINDI

27 June 1969.

MESSAGE FROM
GENERAL A.M. YAHYA KHAN,
PRESIDENT OF PAKISTAN.

----

Greetings and felicitations from Pakistan to the American Astronauts who blazed a new trail for mankind by landing on the Moon. May their high venture and pioneering courage open a new era of peace and progress for the human race.

(A.M. Yahya Khan)

Greeting and felicitations from Pakistan to the American Astronauts who blazed a new trail for mankind by landing on the Moon. May their high venture and pioneering courage open a new era of peace and progress for the human race.

                                                A.M. Yahya Kahn
                                                President

# Panama

```
RCA
                                                         NASA
    YW1537 PAN486 PRO3
    URWA CO PAPA 042                            1969 JUN 25 22 27Z
    PANAMARP CK42 PANGVT 25 1633 VIATROPICAL/RCA


ETAT
SENOR
T. O. PAINE
ADMINISTRADOR NATIONAL AERONAUTICS AND SPACE ADMINISTRATION
WASHINGTONDC



HERE MEN FROM PLANET EARTH FIRST SET FOOT
UPON THE MOON WE COME IN PEACE FOR ALL
MANKIND
        CORONEL BOLIVAR URRUTIA P.
        PRESIDENTE ENCARGADO DE LA JUNTA PROVISIONAL DE
        GOBIERNO
```

All nations of the Earth, small and large, share the wish that the arrival of the first men on the Moon will be a permanent message of peace. Panama is among the first to make this fervent wish of mankind its own.

Colonel Bolivar Urrutia P.
President

# Peru

> PERUVIAN EMBASSY
> WASHINGTON 6, D.C.
>
> El Gobierno y el pueblo del Perú se asocian a los astronautas del Apolo 11 en su extraordinario viaje a la Luna y expresan su ferviente anhelo de que las inmensas posibilidades del espíritu humano que ha realizado la conquista del espacio sea igualmente capaz de asegurar entre las Naciones una era de Paz y de Justicia.
>
> General Juan Velasco Alvarado
> Presidente de la República Peruana

The Government and the people of Peru join in spirit the astronauts of Apollo 11 in their extraordinary trip to the Moon and express their fervent wish that the immense possibilities of the human spirit which have conquered space may be equally capable of insuring among the Nations of the Earth an era of peace and justice.

**General Juan Velasco Alvarado**
President

# Philippines

**Office of the President**
of the Philippines

MESSAGE

The age-old dream of man to cut his bonds to Planet Earth and reach for the stars has given him not only wings, but also the intellect and the intrepid spirit which had enabled him to overcome formidable barriers and accomplish extraordinary feats in the exploration of the unknown, culminating in this epochal landing on the moon.

The Filipino people join the world in congratulating the United States of America for putting the first men on the moon, a triumphant milestone in the conquest of space which augurs greater achievements in the broadening of man's vision and the fulfillment of a larger destiny, within the framework of true human brotherhood and an enduring peace.

President of the Philippines

1969

The age-old dream of man to cut his bonds to Planet Earth and reach for the stars has given him not only wings, but also the intellect and the intrepid spirit which had enabled him to overcome formidable barriers and accomplish extraordinary feats in the exploration of the unknown, culminating in this epochal landing on the Moon.

The Filipino people join the world in congratulating the United States of America for putting the first men on the Moon, a triumphant milestone in the conquest of space which augurs greater achievements in the broadening of man's vision and the fulfillment of a larger destiny, within the framework of true human brotherhood and an enduring peace.

Ferdinand Marcos
President

# Portugal

Presidência da República

Os portugueses, que nos séculos passados descobriram a terra desconhecida, sabem admirar aqueles que nos nossos dias exploram os espaços intersiderais e conquistam para a Humanidade novos mundos.

Lisboa, 27 de Junho de 1969

*Américo Deus Rodrigues Thomaz*
Américo Deus Rodrigues Thomaz
Presidente da República Portuguesa

The Portugese people, discoverers of the unknown Earth in centuries past, know how to admire those who in our days explore outer space bringing mankind in contact with other worlds.

**Americo Deus Rodrigues Thomaz**
**President**

# Romania

Președintele Consiliului de Stat
al
Republicii Socialiste România

FIE CA PRIMUL CONTACT DIRECT AL OMULUI CU LUNA SA CONTRIBUIE LA INFAPTUIREA ASPIRATIILOR DE PROGRES SI PACE ALE TUTUROR OAMENILOR DE PE PAMINT.

NICOLAE CEAUSESCU

May the first direct contact of man with the Moon contribute to the fulfillment of the aspirations for progress and peace of all people on Earth.

Nicolae Ceausescu
President

# Senegal

DEPARTMENT OF STATE

July 1, 1969

The following message received by telegram
from the President of Senegal:

Ceci est un message des militants de la négritude.
C'est un message de solidarité humaine, un message de
paix. Dans cette première visite à la lune, nous saluons
moins une victoire de la technologie qu'une victoire de
la volonté humaine: volonté de recherche et de progrès,
mais aussi de fraternité.

Léopold Sédar Senghor

This is a message from black militants. It is a message of human solidarity, a message of peace. In this first visit to the Moon, rather than a victory of technology we salute a victory of human will: research and progress, but also brotherhood.

Leopold Sedar Senghor
President

# Sierra Leone

Ref: WDC/AD/1/36

7th July, 1969

Mr. T. O. Paine,
Administrator,
National Aeronautics And Space Administration,
Washington, D. C. 20546

Dear Sir,

I have the honour to refer to your letter of 23rd June, 1969 and to transmit herewith the following message sent by the Honourable Prime Minister of Sierra Leone in connection with your request.

"I send the Apollo 11 Astronauts very best wishes for a successful landing on the moon one of the greatest acheivements and triumph of man. Their tasks in connection with this great event are extremely complex and difficult. But a successful completion of this assignment will be recorded as a remarkable landmark in the development of mankind and a victory over the forces of nature. We wish them a pleasant and successful journey and a safe return home."

I am very sorry that I have to transmit this message after June 30th, your deadline. I however, hope that you would be able to make use of it the best way possible.

Yours sincerely,

Victor E. Sumner,
Charge' d'Affaires a. i.

VES/ib

I send the Apollo 11 astronauts very best wishes for a successful landing on the Moon, one of the greatest achievements and triumphs of man. Their tasks in connection with this great event are extremely complex and difficult. But a successful completion of this assignment will be recorded as a remarkable landmark in the development of mankind and a victory over the forces of nature. We wish them a pleasant and successful journey and a safe return home.

Siaka P. Stevens
Prime Minister

# South Africa

Message of the State President of the Republic of South Africa, Mr. Johannes Jacobus Fouché, on the occasion of the first landing on the moon by the Apollo astronauts of the United States of America in July, 1969.

Boodskap van die Staatspresident van die Republiek van Suid-Afrika, mnr. Johannes Jacobus Fouché, by geleentheid van die eerste landing op die maan van die Apollo ruimtevaarders van die Verenigde State van Amerika in Julie 1969.

No human being can be unmoved by the prospect unveiled, at this historic moment in time, of man's first landing on the moon. Thus man reaches out beyond the confines of his own planet, in an enterprise in which the United States of America and its heroic astronauts have opened a new dimension to human experience. I am proud of South Africa's association from the outset with NASA's space programme. On behalf of the Government and all the peoples of the Republic of South Africa, I salute this manifestation of human courage and enterprise and express heartfelt wishes and our prayers to the Almighty for the success of this climactic project.

Niemand kan op hierdie historiese oomblik afsydig staan teenoor die vooruitsigte wat deur die mens se eerste landing op die maan onthul word nie. Op hierdie wyse reik die mens buite die grense van sy eie planeet, in 'n onderneming waarin die Verenigde State van Amerika en sy heldhaftige ruimtevaarders 'n nuwe dimensie vir menslike ondervinding geopen het. Ek is trots op Suid-Afrika se verbintenis van die staanspoor af met NASA se ruimteprogram. Namens die Regering en alle volkere van die Republiek van Suid-Afrika, huldig ek hierdie manifestasie van menslike dapperheid en ondernemingsgees en wil my opregte wense en ons bede tot die Almagtige uitspreek vir die sukses van hierdie spitsprojek.

No human being can be unmoved by the prospect unveiled, at this historic moment in time, of man's first landing on the Moon. Thus man reaches out beyond the confines of his own planet, in an enterprise in which the United States of America and its heroic astronauts have opened a new dimension to human experience. I am proud of South Africa's association from the outset with NASA's space program. On behalf of the Government and all the peoples of the Republic of South Africa, I salute this manifestation of human courage and enterprise and express heartfelt wishes and our prayers to the Almighty for the success of this climactic project.

**Johannes Jacobus Fouche**
**President**

# Swaziland

**Man steps upon the Moon with pride, faith and hope as his inspiration and peace with progress as his objective.**

**H.M. King Sobhuza II**

# Thailand

# Apollo 11 Goodwill Messages

The Thai people rejoice in and support this historic achievement of Earth men, as a step towards Universal peace.

Bhomuibol Adulyadej
King of Thailand

## 222 We Came In Peace For All Mankind

# Togo

AMBASSADE DU TOGO
WASHINGTON, D. C.

Le Général Etienne EYADEMA, Président de la

République Togolaise, au nom du peuple Togolais, s'associe

aux Etats-Unis d'Amérique et d'autres nations de la Terre

pour porter à la Lune un voeu de PAIX, une PAIX UNIVERSELLE,

SEULE ET INDIVISIBLE.

Pour le Président de la République et par ordre

L'Ambassadeur du Togo
Aux Etats-Unis

Alexandre J. OHIN, M.D., F.A.C.S.

Washington, le 30 Juin 1969

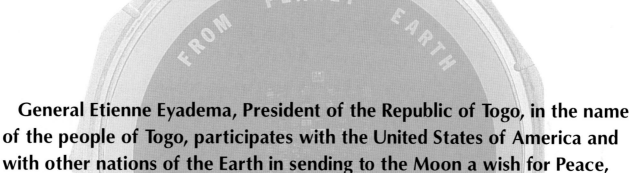

General Etienne Eyadema, President of the Republic of Togo, in the name of the people of Togo, participates with the United States of America and with other nations of the Earth in sending to the Moon a wish for Peace, Universal Peace, Unique and Indivisible. For and by the order of the President of the Republic.

                                         **General Etienne Eyadema**

# Trinidad & Tobago

> The Government and people of Trinidad and Tobago acclaim this historic triumph of science and the human will. It is our earnest hope for mankind that while we gain the moon, we shall not lose the world.
>
> Eric Williams
> Prime Minister
> Trinidad and Tobago

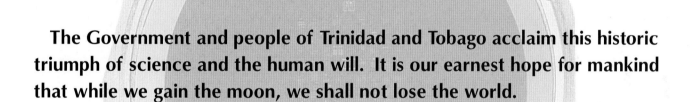

The Government and people of Trinidad and Tobago acclaim this historic triumph of science and the human will. It is our earnest hope for mankind that while we gain the moon, we shall not lose the world.

Eric Williams
Prime Minister

# Tunisia

LE TEXTE ORIGINAL

"Une nouvelle page de l'histoire universelle commence aujourd'hui: celle de la conquete de la lune par l'homme. L'épopée humaine s'enrichit ainsi d'une grande prouesse scientifique et un nouvel horizon s'ouvre pour les habitants de notre planete grace notamment a la victoire des techniciens et des astronautes Americains.

Je souhaite que cette date memorable entre toutes préfigure une ere de paix et de concorde entre tous les hommes, unis dans une même volonté d'oeuvrer pour le triomphe du génie pacifique et createur de l'humanité."

Habib Bourguiba
President de la République Tunisienne

A new page in the history of the Universe begins today: that of man's conquest of the moon. The human epic is enriched by great scientific prowess and a new horizon is opened to the inhabitants of our planet thanks, particularly, to the success of the American technicians and astronauts.

I hope that this date, memorable to all, signifies an era of peace and accord among all men, allied to strive for the triumph of the peaceful and creative mind of humanity.

**Habib Bourguiba**
**President**

# Turkey

Ankara, Temmuz 1969.

İnsanların Ay'a inişini, bugüne kadar ancak tahayyül edebildiğimiz yeni bir çağın başladığına işaret sayıyorum.

Türk milleti, binlerce yıllık bu rüyanın gerçekleşmesini, feza seferlerine girişildiğinden beri hararetle temenni etmiş ve bu yolda sağlanan başarıları heyecan ve ümitle izlemiştir.

Bugünkü insanlığın ve geliştirdiği medeniyetin, barışçı maksatlar için değerlendireceğine inandığımız bu fevkal'ade sonuca ulaşmasında emeği geçenleri, kahraman astronotları ve hiçbir fedakârlığı esirgemeyen dost ve müttefikimiz Amerikan milletini yürekten tebrik ederim.

Cevdet Sunay
Türkiye Cumhurbaşkanı

I consider the landing of the men on the moon as a sign of the beginning of a new era of which we could hardly dream until now.

Since the start of the space explorations, the Turkish nation has most ardently wished the realization of this thousand-year old dream and followed with great hope and excitement every success in this field.

I wish to congratulate the most heartily the heroic astronauts and the American people, our friends and allies, for they have spared no effort in this field and also those who have contributed to the achievement of this outstanding accomplishment from which, I am sure, mankind and our civilization will benefit for peaceful purposes.

Cevdet Sunay
President

# Upper Volta

Message from His Excellency General Sangoule Lamizana,
President of the Republic of Upper Volta

At a time when, transcending the terrestrial sphere and overcoming distances, man is about to dispel the mystery enveloping this world by conquering the Moon, I, General Sangoule Lamizana, President of the Republic of Upper Volta, invite men of the entire earth to unite to celebrate this victory which does not belong only to one country or one people but simply to the glory of all mankind.

We hope with all our heart that this great exploit which we salute with respect and admiration will bring mankind progress and happiness in all areas and that men, feeling the universe become smaller, will strive to make the earth become henceforth a common fatherland where the sons of Adam will live united in brotherhood.

General Sangoule Lamizana

At a time when, transcending the terrestrial sphere and overcoming distances, man is about to dispel the mystery enveloping this world by conquering the Moon, I General Sangoule Lamizana, President of the Republic of Upper Volta, invite men of the entire earth to unite to celebrate this victory which does not belong only to one country or one people but simply to the glory of all mankind.

We hope with all our heart that this great exploit which we salute with respect and admiration will bring mankind progress and happiness in all areas and that men, feeling the universe become smaller, will strive to make the earth become henceforth a common fatherland where the sons of Adam will live united in brotherhood.

**General Sangoule Lamizana
President**

# Uruguay

Presidencia de la República Oriental del Uruguay

Montevideo, junio 27 de 1969.

Señor Presidente:

    El Gobierno y el pueblo del Uruguay han seguido con vivísimo interés la maravillosa y continuada crónica de la conquista del espacio exterior que culminará en fecha muy próxima, dentro de pocos días, con la llegada a la Luna de una aeronave americana tripulada.

    Al hacerme intérprete de tales sentimientos, he querido enviarle, Excelentísimo señor Presidente, mis más cordiales felicitaciones por esta hazaña, que permanecerá como uno de los hechos más salientes de nuestro siglo, fruto del coraje, de la tenacidad y del más inquebrantable deseo de superación.

    Las generaciones venideras recibirán los frutos positivos de este esfuerzo, que beneficia al entero universo.

    Reciba usted Excelentísimo señor Presidente, mis más cordiales y amistosos saludos.

Al Excelentísimo
Señor Presidente de los Estados Unidos de América
Don Richard Nixon.

The Government and the people of Uruguay have followed with great interest the marvelous and continuing chronicle of the conquest of outer space which will culminate very soon, within a few days, with the arrival on the Moon of a manned American spaceship.

In order to express my feelings on this matter, I wish to send you, Mr. President, my most cordial congratulations on this heroic feat, which will remain one of the most outstanding achievements of our century, the fruit of courage, tenacity, and the most unshakable desire to excel.

Future generations will enjoy the positive results of this endeavor, which will benefit the entire universe. Best regards.

Jorge Pacheco-Areco
President

# Vatican

*Jahweh our Lord, how great your name throughout the earth, above the heavens is your majesty chanted.*

*By the mouths of children, babes in arms, you set your stronghold firm against your foes to subdue enemies and rebels.*

*I look up at your heavens, made by your fingers, at the moon and stars you set in place.*

*Ah, what is man that you should spare a thought for him? Or the son of man that you should care for him?*

*You have made him a little less than an angel, you have crowned him with glory and splendor, and you have made him lord over the work of your hand.*

*You set all things under his feet, sheep and oxen all these, yes, wild animals too, birds in the air, fish in the sea traveling the paths of the ocean.*

*Jahweh our Lord, how great your name throughout the earth!*

**Psalms 8**

*To the glory of the name of God who gives such power to men, we ardently pray for this wonderful beginning.*

*Paul VI, Pope*

# Vietnam

```
              MESSAGE OF GOOD WILL
    FROM THE PRESIDENT OF THE REPUBLIC OF VIET NAM,
    H.E. NGUYEN VAN THIEU, FOR THE APOLLO 11
         ASTRONAUTS TO DEPOSIT ON THE MOON
                  JULY 21, 1969

                       -o-


        For many thousand years, the moon has been celebrated
by the Vietnamese poets as a beautiful paradise. Today we
already know that the face of the moon does not correspond
to the imagination of ancient poets. The fact, however, that
men of the earth finally set foot on the moon marks the
beginning of a most beautiful adventure, because it opens
broader vistas on our immense universe, and the perspectives
on men's accessibility to other worlds.


        This memorable feat should bring to mankind both
a sense of pride anh humility: pride, because human beings
by their intelligence and perseverance are now able to
get beyond this earth to which they seemed to be bound;
humility, because the quarrels which divide men on the
earth look so petty in the context of our vast universe.


        In this historical event, we prayerfully hope
that the ingenuity and intelligence which God endows to
men will lead toward increasing progress and brotherhood,
and the widening of human horizons. We are, therefore
very happy that the first message deposited by the brave
american astronauts of Apollo 11 on the moon is a message
of peace for all mankind, and from all mankind, in which
the Vietnamese people fully concur.
```

For many thousand years, the moon has been celebrated by the Vietnamese poets as a beautiful paradise. Today we already know that the face of the moon does not correspond to the imagination of ancient poets. The fact, however, that men of the earth finally set foot on the moon marks the beginning of a most beautiful adventure, because it opens broader vistas on our immense universe, and the perspectives on men's accessibility to the other worlds.

This memorable feat should bring to mankind both a sense of pride and humility: pride, because human beings by their intelligence and perseverance are now able to get beyond this earth to which they seemed to be bound; humility, because the quarrels which divide men on the earth look so petty in the context of our vast universe.

In this historical event, we prayerfully hope that the ingenuity and intelligence which God endows to men will lead toward increasing progress and brotherhood, and the widening of human horizons. We are, therefore, very happy that the first message deposited by the brave American astronauts of Apollo 11 on the moon is a message of peace for all mankind, and from all mankind, in which the Vietnamese people fully concur.

> H.E. Nguyen Van Thieu
> **President**

# Yugoslavia

> YUGOSLAV EMBASSY
> WASHINGTON
>
> PORUKA DOBRE VOLJE
> PREDSEDNIKA TITA KOJU ĆE ASTRONAUTI
> APOLO 11 POLOŽITI NA MESEC
>
> "Neka ovo grandiozno ispunjenje drevnog
> sna ljudskog roda - stupanje na daleko tlo mjeseca,
> tog prvog susjeda svih nas - približi ostvarenje
> vjekovne težnje čovječanstva da živi u miru,
> bratstvu i saradnji.
>
> Josip Broz Tito
> Predsednik Socijalističke
> Federativne Republike Jugoslavije"

May this majestic fulfillment of the ancient dream of the human race — man's setting foot on the distant soil of the moon, the first neighbor of us all — bring closer the realization of humanity's age-long vision to live in peace, brotherhood and joint endeavor.

Josip Broz Tito
President

# Zambia

EMBASSY OF THE REPUBLIC OF ZAMBIA
1875 CONNECTICUT AVENUE, N.W.  WASHINGTON, D.C. 20009
TELEPHONE: 265-9717  CABLES: ZAMBIAN

June 25, 1969

Mr. T. O. Paine
Administrator
National Aeronautics & Space Administration
Washington, D.C. 20546

Dear Mr. Paine:

    With reference to your letter dated June 23, 1969, we have received the following message from our Government to be taken to the moon by the Apollo astronauts:

"ZAMBIA SENDS FROM EARTH THROUGH EARTH'S FIRST MESSENGERS TO LAND ON THE MOON - THE INTREPID ASTRONAUTS ARMSTRONG, ALDRIN AND COLLINS - THIS MESSAGE OF GOOD WILL AMONG MEN OF ALL LANDS AND CITIES: THAT THERE BE HARMONY IN ALL THE CREATED WORLD STOP"

Sincerely yours,

For: Chargé d'Affaires a.i.

PKM:p

## Apollo 11 Goodwill Messages

Zambia sends from Earth through Earth's first messengers to land on the moon — the intrepid astronauts Armstrong, Aldrin and Collins — this message of good will among men of all lands and cities: that there be harmony in all the created world.

**Kenneth David Kaunda**

# Bolivia

```
QUOTE CON FE INQUEBRANTABLE EN LOS POSTULADOS DE
LA PAZ, DE LA JUSTICIA Y DEL PROGRESSO EN LA ETAPA
HISTORICA QUE SE INICIA EL 16 DE JULIO DE 1969, EL
PUEBLO DE BOLIVIA SALUDA LA HAZANA DEL APOLO 11, Y
EL PROGRAMA ESPACIAL DE LOS ESTADOS UNIDOS DE
AMERICA QUE HACE POSIBLE EL ARRIBO DEL HOMBRE A LA
LUNA COMO SIMBOLO DE SU GENIO Y CORAJE. DR. LUIS
ADOLFO SILES SALINAS, PRESIDENTE CONSTITUCIONAL DE
LA REPUBLICA. LA PAZ, JULIO DE 1969 END QUOTE
CASTRO
```

With unshakable faith in the postulates of peace, justice, and progress in the historic period berginning July 16, 1969, the people of Bolivia salute the exploit of Apollo 11, and the space program of the United States of America which is making possible the arrival of man on the moon as a symbol of his genius and courage. Dr. Luis Adolfo Siles Salinas, Constitutional President of the Republic. La Paz, July, 1969

Other Nations 245

**Embassy of Ceylon**
2148 WYOMING AVENUE, N.W.
WASHINGTON, D.C. 20008

IN REPLY, PLEASE
QUOTE REFERENCE NO    POL/
PHONE 202 HU 3-4025

15 July, 1969

Dear Mr. Paine,

I write with reference to your letter of June 23 to His Excellency the Ambassador by which you were kind enough to inform that NASA was prepared to take a message from the Prime Minister of Ceylon for depositing on the moon.

The Government of Ceylon whilst thanking NASA for its kindness in requesting such a message has decided not to send such a message.

I should like on behalf of His Excellency the Ambassador to express thanks for your interest in this matter, and for the cordial invitation that was extended to us to send a message.

Yours sincerely,

A. T. Jayakoddy,
Charge d'Affaires a.i.

Mr. T. O. Paine,
Administrator,
National Aeronautics and Space
            Administration,
Washington, D.C. 20546.

atj/ymf

# Finland

EMBASSY OF FINLAND
1900 TWENTY-FOURTH ST., N.W.
WASHINGTON, D.C. 20008

1923

June 23, 1969

Mr. T. O. Paine
Office of The Administrator
National Aeronautics and Space Administration
Washington, D. C. 20546

Dear Mr. Paine:

    Thank you for your letter of June 23, 1969, concerning the messages that will be carried to the moon by the Apollo 11 astronauts.

    We regret, however, that it will not be possible to complete the necessary communications to Finland by the stated deadline of June 30, 1969.

    We greatly appreciate your notice to the Embassy on this matter.

Sincerely yours,

Lauri Hannikainen
Attaché

LH/sbd

# Gabon

Message from His Excellency
Albert-Bernard Bongo
President of the Republic of Gabon

When the Apollo 11 astronauts hurl into space to place men on another planet for the first time, all my encouragement and that of the people of Gabon accompanys them in this exciting adventure.

This historic voyage, marked by the seal of peace and brotherhood, does honor not only to the American people but to all mankind.

(signed) Albert-Bernard Bongo

President of the Republic of Gabon

# Germany

THE GERMAN AMBASSADOR  WASHINGTON, D.C.
July 1, 1969

Dear Dr. Paine:

This is in reference to your kind letter of June 23, 1969, in which you indicated the opportunity to have a good will message by the German President deposited on the moon together with similar messages of other Chiefs of State.

I have been in constant communication with my Government in order to meet the deadline of June 30 as mentioned in your letter. Due to the change of the Presidency in Germany, however, which took place on July first, it was unfortunately not possible to make the administrative and technical arrangements necessary for transmitting the message to Washington in time.

I regret, therefore, not to be in a position to follow your suggestion which deeply impressed me.

Sincerely yours,

Rolf Pauls

Mr. T. O. Paine,
Administrator
National Aeronautics and
Space Administration
Washington, D.C. 20546

# Poland

Although we are not suggesting any message from the Polish Head of State, please be assured that the achievements of the U.S. astronauts are followed by us with great interest, appreciation and best wishes for the success in their endeavor.

Sincerely,

Jorzy Michalowski
Ambassador

# Sweden

ROYAL SWEDISH EMBASSY

Washington, D C, June 26, 1969

Mr T O Paine, Administrator
National Aeronautics and
Space Administration
WASHINGTON, D C 20546

Dear Mr Paine:

In Ambassador de Besche's absence I wish to acknowledge the receipt of your letter of June 23 with your kind offer to accept a message from the Swedish Chief of State to be deposited on the moon. I have forwarded the contents of your letter to my Government. In case the King of Sweden avails himself of this opportunity to have a message taken to the moon we will see to it that the message is received by you before June 30.

Sincerely yours,

P B Kollberg
Chargé d'Affaires of Sweden

# ETCHED ON APOLLO 11 DISC

## PRESIDENTIAL STATEMENTS

"...The Congress hereby declares that it is the policy of the United States that activities in space should be devoted to peaceful purposes for the benefit of all mankind..."

> National Aeronautics and Space
> Act of 1958
> Signed by President
> Dwight D. Eisenhower
> July 29, 1958

"...We go into space because whatever mankind must undertake, free men must fully share. ...I believe that this Nation should commit itself to achieving the goal before this decade is out, of landing a man on the moon and returning him safely to earth. No single space project in this period will be more exciting, or more impressive to mankind, or more important for the long-range exploration of space; and none will be so difficult or expensive to accomplish..."

> President John F. Kennedy
> May 25, 1961

"...We expect to explore the moon, not just visit it or photograph it. We plan to explore and chart planets as well. We shall expand our earth laboratories into space laboratories and extend our national strength into the space dimension. The purpose of the American people – expressed in the earliest days of the Space Age – remains unchanged and unwavering. We are determined that space shall be an avenue toward peace and we both invite and welcome all men to join with us in this great opportunity..."

> President Lyndon B. Johnson
> January 27, 1965

"...Our current exploration of space makes the point vividly: Here is testimony to man's vision and to man's courage. The journey of the astronauts is more than a technical achievement; it is a reaching-out of the human spirit. It lifts our sights; it demonstrates that magnificent conceptions can be made real. They inspire us and at the same time they teach us true humility. What could bring home to us more the limitations of the human scale than the hauntingly beautiful picture of our earth seen from the moon? ..."

> President Richard M. Nixon
> June 4, 1969

# ETCHED ON APOLLO 11 DISC

## THE UNITED STATES HOUSE OF REPRESENTATIVES

George P. Miller, Chairman

Olin E. Teague
Joseph E. Karth
Ken Hechler
Emilio Q. Daddario
John W. Davis
Thomas N. Downing
Joe D. Waggonner, Jr.
Don Fuqua
George E. Brown, Jr.
Earle Cabell
Bertram L. Podell
Wayne N. Aspinall
Roy A. Taylor
Henry Helstoski
Mario Biaggi
James W. Symington
Edward I. Koch

James G. Fulton
Charles A. Mosher
Richard L. Roudebush
Alphonzo Bell
Thomas M. Pelly
John W. Wydler
Guy Vander Jagt
Larry Winn, Jr.
Jerry L. Pettis
Donald E. Lukens
Robert Price
Lowell P. Weicker, Jr.
Louis Frey, Jr.
Barry Goldwater, Jr.

## Committee on Appropriations

George H. Mahon, Chairman        Frank T. Bow

## Subcommittee on Independent Offices and Department of Housing and Urban Development

Joe L. Evins, Chairman
Edward P. Boland
George E. Shipley

Charles R. Jonas
Louis C. Wyman

Robert N. Giaimo  
John O. Marsh, Jr.  

Burt L. Talcott  
David H. Pryor  

## THE UNITED STATES SENATE

Spiro T. Agnew  
President of the Senate

Richard B. Russell, President pro tempore  
Michael J. Mansfield  
Edward M. Kennedy  

Everett McKinley Dirksen  
Hugh Scott  

### Committee on Aeronautical and Space Sciences

Clinton P. Anderson, Chairman  
Richard B. Russell  
Warren G. Magnuson  
Stuart Symington  
John Stennis  
Stephen M. Young  
Thomas J. Dodd  
Howard W. Cannon  
Spessard L. Holland  

Margaret Chase Smith  
Carl T. Curtis  
Mark O. Hatfield  
Barry Goldwater  
Charles McC. Mathias, Jr.  
William B. Saxbe  

### Committee on Appropriations

Richard B. Russell, Chairman  

Milton R. Young

## Subcommittee on Independent Offices and Department Of Housing and Urban Development

John O. Pastore, Chairman
Warren G. Magnuson
Allen J. Ellender
Richard B. Russell
Spesssard L. Holland
John Stennis
Michael J. Mansfield

Gordon Allott
Margaret Chase Smith
Roman L. Hruska
Norris Cotton
Clifford P. Case

## NATIONAL AERONAUTICS AND SPACE ADMINISTRATION OFFICIALS

| | |
|---|---|
| T. Keith Glennan | Administrator 1958-1961 |
| Hugh L. Dryden | Deputy Administrator 1958-1965 |
| James E. Webb | Administrator 1961-1968 |
| Robert C. Seamans, Jr. | Deputy Administrator 1966-1967 |

---

| | |
|---|---|
| Thomas O. Paine | Administrator |
| Homer E. Newell | Associate Administrator |
| Willis H. Shapley | Associate Deputy Administrator |
| George E. Mueller | Associate Administrator for Manned Space Flight |
| John E. Naugle | Associate Administrator for Space Science and Applications |

| | |
|---|---|
| Bruce T. Lundin | Associate Administrator for Advanced Research and Technology (Acting) |
| Gerald M. Truszynski | Associate Administrator for Tracking And Data Acquisition |
| Lt. Gen. Samuel C. Phillips | Apollo Program Director |
| Robert R. Gilruth | Director, Manned Spacecraft Center |
| Wernher von Braun | Director, George C. Marshall Space Flight Center |
| Kurt H. Debus | Director, Kennedy Space Center |
| John F. Clark | Director, Goddard Space Flight Center |
| Dr. William H. Pickering | Director, Jet Propulsion Laboratory |
| Robert L. Krieger | Director, Wallops Station |
| Edgar M. Cortright | Director, Langley Research Center |
| Abe Silverstein | Director, Lewis Research Center |
| Hans M. Mark | Director, Ames Research Center |
| James C. Elms | Director, Electronics Research Center |
| Paul F. Bikle | Director, Flight Research Center |

# Appendix 1

## List of Leaders represented on the Apollo 11 Silicon Disc

### Afghanistan
Mohammed Zahir Shah (1914 – 2007), was the last King (Shah) of Afghanistan reigned from 1933 to 1973. He lived in exile for 29 years and returned in 2002 during American occupation.

### Argentina
Juan Carlos Onganía Carballo (1914 – 1995), was the President of Argentina from 1966 to 1970. He rose to power in a coup and became a military dictator. He was responsible for La Noche de los Bastones Largos (where he ordered police to invade the University of Buenos Aires where students and professors were beaten).

### Australia
Sir John Grey Gorton (1911 – 2002), was the 19th Prime Minister of Australia from 1968 to 1971. He was elected to the Senate in 1949 as a senator for the state of Victoria, Menzies ministry as Minister for the Navy from 1958 to 1963, Minister for Works from 1963 to 1967, Minister for the Interior from 1963 to 1964, Minister for Education and Science from 1966 to 1968. Gorton became the leader of the Liberal Party and later resigned from that party in 1975.

### Belgium
Baudouin I (1930 – 1993) (French: Baudouin Albert Charles Léopold Axel Marie Gustave or Dutch: Boudewijn Albert Karel Leopold Axel Marie Gustaaf), reigned as King of the Belgians from 1951 to 1993. He was the eldest son of King Leopold III and his first wife, Princess Astrid of Sweden. He brought stability between the Dutch-speaking Flanders and the French-speaking Wallonia. His death brought Flemings and Walloons together in support of the monarchy.

## Brazil
Artur da Costa e Silva (1902 – 1969), was President from 1966 to 1969. He faced strong opposition from his nation's people. He greatly increased the power of the president and closed the Congress. He banned the opposition and increased media censorship. He suffered a stroke in August 1969, just a month after the Apollo 11 goodwill message was sent.

## Canada
Pierre Elliott Trudeau (1919 – 2000), was the 15th Prime Minister of Canada from 1968 to 1979 and 1980 to 1984. He was a charismatic figure, often controversial and had a flamboyant personality. He established the Charter of Rights and Freedoms which redefined Canada.

## Chad
Francois Tombalbaye (1918 – 1975), was President from 1960 to 1975 and established dictatorial control. The Arab north opposed his policies which lead to open rebellion in 1965. In 1975 he was deposed and killed by a military takeover.

## Chile
Eduardo Nicanor Frei Montalva (1911 – 1982), was President from 1964 to 1970. The CIA supported Frei in the election of 1964 and he brought about many reforms during his presidency. He wrote a historical letter to Mariano Rumor (President of the International Christian Democrats), endorsing military intervention and denouncing Communism. He later opposed a Pinochet dictatorship.

## China, Republic of
Chiang Kai-shek (1887 – 1975), was President from 1928 to 1975. He unified the warlords in 1928 as the leader of the Republic of China. He was also the leader during the Second Sino-Japanese War. He tried to defeat Communists in the Chinese Civil War from 1927 to 1950, but failed and retreated to Taiwan. Chiang attempted to eradicate the Chinese Communists, but ultimately failed, forcing his government to retreat to Taiwan where he was President of the Republic of China.

## *Colombia*
Carlos Lleras Restrepo (1908 – 1994), was President of Colombia from 1966 to 1970.  In the late 30's he was Minister of Finance (Ministro de Hacienda).  His Colombian Institute for Agrarian Reform promoted the redistribution of land to the peasants.  He was the founder of the magazine Nueva Frontera.

## *Congo (Zaire)*
Mobutu Sese Seko Nkuku Ngbendu wa Za Banga (1930 – 1997), also called Joseph-Désiré Mobutu, was the President of Zaire (Democratic Republic of the Congo) from 1965 to 1997.  In a bloodless coup, he seized power from President Kasavubu, abolishing the parliament and assuming wide-ranging powers.  An attempted coup against him resulted in the public hanging of four cabinet members. Tutsis had long opposed Mobutu due to his support for Rwandan Hutu extremists.  Rwanda and Uganda leaders joined to overthrow Mobutu.  In 1997, the Tutsi rebels and other groups captured Kinshasa (the capital and largest city) located on the Congo River.  Zaire was renamed the Democratic Republic of Congo and Mobutu was exiled.

## *Costa Rica*
José Joaquín Antonio Trejos Fernández (b. 1916), was President of Costa Rica from 1966 to 1970.  He defeated Daniel Oduber in the election that secured him the presidency.

## *Cyprus*
Makarios III, born Mihalis Christodoulou Mouskos (1913 – 1977), was President of the Republic of Cyprus from 1960 to 1977.  He was also the archbishop and primate of the autocephalous Cypriot Orthodox Church from 1950 to 1977.  He led the movement for enosis (union with Greece), but was exiled by the British in 1956 on charges of encouraging terrorism.  He was released in 1957.  When an agreement was reached on the independence of Cyprus from Great Britain, he was elected president.  He was reelected in 1968 and 1973.  He was overthrown by General George Grivas, leader of the enosis movement in 1974.  After exile, he returned to Cyprus and resumed his presidency.

## *Dahomey (Benin)*
Emile Derlin Zinsou (b. 1918), was President of Dahomey from 1968 to 1969 when he took power in a military regime.  He was the Foreign Minister of Dahomey from 1962 to 1963.  He later opposed Marxist policies of Mathieu Kérékou.  Dahomey (formerly part of French West Africa) is now called the Republic of Benin.

## Denmark
Frederik IX (Christian Frederik Franz Michael Carl Valdemar Georg) (1899 – 1972), was King of Denmark from 1947 to 1972. He was the son of King Christian X of Denmark and Queen Alexandrine, born Duchess of Mecklenburg. Unprecedented changed in Denmark's economy and work force allowed it to become a modern country. He married Princess Ingrid of Sweden. The King had no sons and three daughters, so Princess Margrethe (his eldest daughter), succeeded as Queen Margrethe II.

## Dominican Republic
Joaquín Antonio Balaguer Ricardo (1906 – 2002), was President of the Dominican Republic from 1960 to 1962, 1966 to 1978, and 1986 to 1996. He was a protégé of Rafael Leonidas Trujillo. He was accused of election fraud and of intimidation of opponents. Despite this, he was a leading figure in the history of the Domincan Republic.

## Ecuador
José Maria Velasco Ibarra (1893 – 1979), was President of Ecuador from 1934 to 1935, 1944 to 1947, 1952 to 1956, 1960 to 1961, and 1968 to 1972. A book entitled CIA Diary discusses the events surrounding his fourth presidency. He was a charismatic figure that promised redemption for his people.

## Estonia
Ernst Jaakson (1905 – 1998), was an Estonian diplomat. He kept diplomatic service for an amazing sixty-nine years. During the 1940's, many Western nations did not recognize the Soviet Union's annexation of Estonia. Estonia regained independence in 1991.

## Ethiopia
Haile Selassie I, "Power of the Trinity" (1892 – 1975), was Emperor of Ethiopia from 1930 to 1974 and defacto Emperor from 1916 to 1936 and 1941 to 1974. Emperor Haile Selassie I spent five years of exile from 1936 to 1941 mainly in the United Kingdom. He is the religious symbol for God incarnate among the Rastafari movement. His original name was Tafari Makonnen. He brought Ethiopia into the mainstream of African politics. He also got Ethiopia into the League of Nations and the United Nations.

## Ghana
Akwasi Amankwaa Afrifa (1936 – 1979), was the Head of the state of Ghana and leader of the military government in 1969 and Chairman of the Presidential Commission from 1969 to 1970. He was elected as a Member of Parliament in 1979 but then executed with two other former heads of state and five other Generals that same year. He was often referred to by his title Okatakyie.

## Great Britain
Elizabeth II (b. 1926) is Queen of sixteen independent states and their overseas territories and dependencies. She is the eldest daughter of George VI and Elizabeth Bowes-Lyon. The Queen and The Duke of Edinburgh have four children; Prince Charles, Princess Anne, Prince Andrew and Prince Edward. She has reigned for forty-six years.

## Greece
Georgios Zoitakis, (1910 – 1996), was General and Regent from 1967 to 1972. General Georgios Zoitakis acted as Regent for the absent King Constantine II who was overthrown in a coup. In 1973, the monarchy was abolished and Georgios Papadopoulos became the President of Greece. Zoitakis was later condemned by a court for high treason.

## Guyana
Linden Forbes Sampson Burnham (1923 – 1985), was Prime Minister of Guyana from 1964 to 1985. He enacted a "National Security Act" giving the police high levels of Power and authority. In 1970, Burnham declared Guyana to be a "co-operative republic." He established strong relations with Cuba and the Soviet Union. Burnham banned all forms of imports and he nationalized of all the foreign owned industries.

## Iceland
Dr. Kristján Eldjárn (1916 – 1982), was President of Iceland from 1968 to 1980. He was awarded a doctorate for his research into pagan burials in Iceland. He taught at Akureyri Grammar School and the College of Navigations in Reykjavík. He was a curator of the National Museum of Iceland in 1945 and its Director in 1947 until the 1968 presidential election.

## India
Indira Priyadarshini Gandhi (1917 – 1984), was Prime Minister of India for three consecutive terms from 1966 to 1977 and a fourth term from 1980 to 1984. Her father was Jawaharlal Nehru. Because of riots and to retain power, Indira Gandhi declared a state of emergency in 1975. She developed close ties to the Soviet Union. She allowed free elections in 1977 and was voted out of office. She later regained her position as Prime Minister. In 1971, India sent its first satellite into space. Economically, she helped India to become one of the fastest growing economies in the world. On October 31, 1984, Indira Gandhi's Sikh bodyguards assassinated her to avenge the storming of the Sikh temple that Gandhi had ordered to stop extremists.

## Iran
Mohammad Reza Pahlavi, Shah of Iran (1919 – 1980), held the imperial titles Shahanshah (King of Kings) and Aryamehr (Light of the Aryans). He was the monarch of Iran from 1941 to 1979. He was the second monarch of the Pahlavi dynasty and the last Shah of the Iranian monarchy. The CIA and British intelligence funded and led a covert operation to depose the Shah's predecessor, Mossadeq with the help of military forces loyal to the Shah, known as Operation Ajax. The Shah westernized the country, received U.S. military aid and recognized Israel, which led to increasing Shiite clergy alienation. Political prisoners were taken into custody by the Shah. In 1979, an Islamic revolution led to the Shah's overthrow.

## Ireland
Éamon de Valera (1882 – 1975), was President of Ireland for two terms from 1959 to 1973. He was a pivotal leader of Ireland's struggle for independence from the United Kingdom of Great Britain and Ireland in the early 20th century and the Irish Civil War. He was the author of Ireland's constitution, Bunreacht na hÉireann. He served three times as Irish head of government; as Príomh Áire, as the second President of the Executive Council and the first Taoiseach.

## Israel
Zalman Shazar, born as Shneur Zalman Rubrashov (1889 – 1974), was President of Israel from 1963 to 1973. Shazar was born to a Hasidic family of the Chabad-Lubavitch near Minsk. He became involved in the Poalei Zion Movement. Shazar settled in Israel in 1924 and became a member of the secretariat of the Histadrut. He was the education minister in David Ben-Gurion's Mapai government from 1949 to 1951, then continuing as a member of Knesset.

## *Italy*
Giuseppe Saragat (1898 – 1988), was President of the Italian Republic from 1964 to 1971. Saragat was a moderate socialist who split from the Italian Socialist Party, because of its alliance with the Communists, to found the Italian Socialist Workers' Party which later became the Italian Democratic Socialist Party. He served as the foreign minister of Italy from 1963 to 1964.

## *Ivory Coast*
Félix Houphouët-Boigny (1905 – 1993), was the first President of Ivory Coast (Côte d'Ivoire) from 1960 to 1993. He was previously a member of the French parliament and appointed minister in the government of France from 1957 to 1961. He maintained a close relationship with France and the West. He maintained an anticommunist foreign policy and stopped diplomatic relations with the Soviet Union and did not recognize China until 1983. After his death, conditions deteriorated.

## *Jamaica*
Hugh Lawson Shearer (1923 – 2004), was Prime Minister of Jamaica from 1967 to 1972. He was a member of the House and the Senate. He remained popular, save for an episode of riots triggered by the banning of an author, Walter Rodney. Shearer did not feel that Rodney's socialist ties to Cuba and the Soviet Union were good for Jamaica. He brought economic growth and educational improvement to his country.

## *Japan*
Eisaku Sat (1901 – 1975), was Prime Minister of Japan, served three terms from 1964 to 1972. He was a popular leader with a growing economy and foreign policy which was good towards China and the United States. He introduced the three non-nuclear principles: non-production, non-possession and non-introduction. He asked President Johnson to return Okinawa. This was later a deal he made with Nixon. The U.S. would be allowed to keep U.S. bases there and Japan would repatriate Okinawa. He opposed the Nixon trip to China and China's introduction into the U.N. Sat shared a Nobel Peace Prize with Seán MacBride in 1974 for the Nuclear Non-Proliferation Treaty.

## *Kenya*
Jomo Kenyatta (1889 – 1978), was Prime Minister from 1963 to 1964 and later President from 1964 to 1978. He was the founding father of Kenya. He pursued a pro-Western, anti-Communist policy. He ruled with an authoritarian style, but was a popular figure. Tribal rivalries made his leadership controversial.

## Korea
Park Chung-hee (1917 – 1979), was President of the Republic of Korea from 1961 to 1979. He gained power in a popular coup as general of the Republic of Korea. He was credited for economic growth, but was criticized for his authoritarian way of ruling the country, for sending troops to help the United States in the Vietnam War and for alleged pro-Japanese activities as a Chinilpa. Park Chung-hee was assassinated in 1979 by Kim Jae-kyu, the director of the intelligence agency. The assassination is considered by some as a spontaneous act of passion by an individual, others think it was as part of a pre-arranged attempted coup.

## Laos
Savang Vatthana or Samdach Brhat Chao Maha Sri Vitha Lan Xang Hom Khao Phra Rajanachakra Lao Parama Sidha Khattiya Suriya Varman Brhat Maha Sri Savangsa Vadhana) (1907 – 1978), was the King of Laos from 1959 to 1975. His rule ended with a takeover of the Pathet Lao in 1975. In 1951, he served as Prime Minister, and in 1959, he was assigned Regent. In 1975 he was forced to abdicate after the communist revolution. He refused to leave the country. In 1977, he was put in an internment camp. The exact year of his death is disputed.

## Latvia
Anatols Dinbergs (1948 – 1993), was Head of Latvian Diplomatic Service from 1971 to 1991. Latvia was occupied by the Soviet Union in 1940. But the United States, from 1940 through 1991 never recognized the forcible annexation of the Baltic States. Latvia was not allowed to establish a government-in-exile in any Western country or sign the Declaration of the United Nations. Latvia's parliament declared full independence on August 21, 1991 in the aftermath of a failed Soviet coup attempt. Anatols Dinbergs was promoted to the rank of Ambassador and Permanent Representative to the United Nations.

## Lebanon
Charles Helou (1913 – 2001), was President of Lebanon from 1964 to 1970. Helou's presidency enjoyed economic growth. Lebanon struggled to avoid involvement in the Arab-Israeli conflict of 1967. Muslims wanted Lebanon to join the Arab war effort while Christians wished to avoid participation. The authority of Lebanon was challenged by the presence of Palestinian guerrillas. There were clashes between the Lebanese army and the Palestine Liberation Organization (PLO). Helou resisted PLO demands, but in 1969, after failing to end the rebellion military, agreed to the Cairo Agreement. This permitted Palestinian guerrillas to launch raids into Israel from bases inside Lebanon.

## Lesotho
Joseph Leabua Jonathan (1914 – 1987), was the first Prime Minister of Lesotho from 1965 to 1970. He was later Tona Kholo (unelected prime minister). Lesotho was economically dependent on South Africa. Jonathan denounced the South African government's policy of apartheid and he declared his support for the prohibited African National Congress. In 1986 a military coup deposed the Jonathan government. Jonathan was placed under house arrest and later died.

## Liberia
William Vacanarat Shadrach Tubman (1895 – 1971), was the President of Liberia from 1944 to 1971. Tubman's administration declared war against Nazi Germany and Japan. Liberia was an important U.S. Ally. The U.S. constructed the Free Port of Monrovia and set up landing strips in Liberia. Tubman maintained pro-Western policies, notably with U.S. President Lyndon B. Johnson. A gunman attempted to assassinate Tubman in 1955 backed by political opponents.

## Madagascar
Philibert Tsiranana (1912 – 1978), was the first President of Madagascar from 1959 – 1972. Tsiranana was elected to the French National Assembly and then helped to form the Social Democratic Party. He became Prime Minister in 1958. When Madagascar gained its independence, the position of Prime Minister was abolished and Tsiranana became Madagascar's first president. Tsiranana resigned amidst protests in 1972.

## Malaysia
Ismail Nasiruddin Shah or Almarhum Sultan Ismail Nasiruddin Shah ibni Almarhum Sultan Zainal Abidin III (1907 – 1979), was Sultan from 1949 to 1979. He was ill while in office and nearly resigned in 1969. He was a very good photographer and entered his works for competition. He won numerous honors and the Grand Prize in a Malayan International Photography Salon in Singapore.

## Maldives
Ibrahim Nasir Rannabandeyri Kilegefan (b. 1926), was President from 1968 to 1978. He was the Prime Minister under Sultan Muhammad Fareer Didi from 1957 to 1968 and succeeded him to become the first President of the Second Republic of Maldives. President Nasir opened the isolated nation to the world. He achieved U.N. Representation and modernized the country. He brought TV and radio to the nation. He also created an English based curriculum for schools.

## Mali

General Moussa Traoré (b. 1936), was President from 1968 to 1991. He became the leader by a coup that deposed President Modibo Keïta. When he seized power, all political activity was banned and a police state was formed. In 1977, ex-president Modibo Keïta died in detention. This aroused suspision about the government. Traoré's government arrested both the defense and security ministers, on accusations of plotting a coup. In 1991, amidst voilence, a military coup removed Traoré and General Amadou Toumani Touré took control. Traoré was condemned to death for alleged crimes and then later pardoned.

## Malta

Giorgio Borg Olivier (1911 – 1980), was the Prime Minister of Malta from 1950 to 1955 and 1962 to 1971. Following the February 1962 election, Dr. Borg Olivier agreed to form a Government after obtaining important amendments to the Constitution. Olivier headed a Government delegation for the Malta Independence Conference that announced the independence of Malta from the British. In 1964, Olivier was made a Knight Grand Cross of the Order of St. Sylvester by Pope Paul VI.

## Mauritius

Sir Seewoosagur Ramgoolan (1900 – 1985), was Prime Minister from 1961 to 1982. An admirer of Gandhi, he led his country to independence from the United Kingdom in 1968. He was Governor-General of Mauritius from 1983 until his death in December 1985. Ramgoolam was made a knight by Queen Elizabeth on June 12, 1965. Some regard him as the father of the nation.

## Mexico

Gustavo Díaz Ordaz Bolaños Cacho (1911 – 1979), was President of Mexico from 1964 to 1970. President Díaz Ordaz was known for his strict, authoritarian manner. The 1968 Summer Olympics, brought protestors and violence. A massacre took place and his actions were criticized as an atrocity. Díaz Ordaz handled the Mexican economy well and industry improved.

## Morocco

King Hassan II (1929 – 1999), was the King of Morocco from 1961 to 1999. The King retained large powers he eventually used to strengthen his rule, which provoked political protest. Elections were often rigged in favor of loyal parties. Protest, demonstrations and riots often challenged the King's rule. In the 1970s, King Hassan survived two assassination attempts, including an attempted coup. Hassan and the CIA had close ties. He served as a back channel for the Arab world and Israel. He regained the Spanish Sahara and Western Sahara after the "Green .March" in 1975. Later, relations with Algeria deteriorated sharply.

## Netherlands

Juliana (Juliana Emma Louise Marie Wilhelmina van Oranje-Nassau) (1909 – 2004), was the queen regnant of the Kingdom of the Netherlands from 1940 to 1980. She reigned after her mother's abdication in 1948. During the Nazi German occupation of the Netherlands, the Prince and Princess decided to leave the Netherlands and moved to the United Kingdom and then to Canada, in exile. In 1945, she returned to the Netherlands. Once home, she expressed her gratitude to Canada by sending the city of Ottawa 100,000 tulip bulbs. Several scandals hit the royal family, but people often admired her compassion. When a destructive storm hit the nation, Queen Juliana waded in water and mud all over the devastated areas to bring people food and clothing.

## New Zealand

Sir Keith Jack Holyoake (1904 – 1983), was the National Party Prime Minister from September 20, 1957 to December 2, 1957 then again from December 12, 1960 to February 7, 1972. He was known for his diplomatic style and as a consensus builder. In 1972 he resigned as Prime Minister to ease the succession for his deputy and friend, Jack Marshall.

## Nicaragua

Anastasio ("Tachito") Somoza Debayle (1925 – 1980), was officially the 44th and 45th President of Nicaragua from 1967 to 1972 and 1974 to 1979. He was the last member of the Somoza family to be President. Somoza launched a violent campaign against the Sandinista Front. The Front was largely backed by Cuba and the Soviet Union. President Jimmy Carter withdrew support for the Somoza regime. In 1979, Somoza resigned the presidency and was denied entry to the U.S. after which he took refuge in Paraguay. He was assassinated there at the age of 54. Some believe that Cuba was responsible for this, since Somoza supported the Bay of Pigs debacle.

## Norway

Olav V (1903 – 1991), was King of Norway from 1957 to 1991. Succeeding to the Norwegian Throne in 1957 (upon the death of Haakon VII), Olav was known as the "People's King". He was immensely popular and often drove his own cars. In 1961 the King was a laureate of the Nansen Refugee Award. Thousands mourned his death in 1991. Norwegians lit hundreds of thousands of candles in the courtyard outside the Royal Castle with letters and cards placed amongst them. Olav's son Harald V succeeded him as King.

## Pakistan
Agha Muhammad Yahya Khan (1917 – 1980), was President of Pakistan from 1969 to 1971. An army general, Yahya Khan declared martial law in the country. When he tried to leave power and conducted an election, trouble started. The political parties from East and West Pakistan refused to recognize the other as a winner. Yahya Khan used military forces in East Pakistan to stop opposition. This resulted in the beginning of the war between Pakistan and India in 1971. War led to the surrender of Pakistani forces to the allied forces of India and Bangladesh. Yahya Khan resigned, to be replaced by Zulfiqar Ali Bhutto. Bhutto placed him under house arrest and Bangladesh became an independent country.

## Panama
Colonel Bolivar Urrutia. In 1968, Colonel Jose Mario Pinilla and Colonel Bolivar Urrutia became heads of a Provisional Junta Government by means of a military coup. Government power was exercised by the National Guard under the leadership of Colonel Omar Torrijas Herrera and Colonel Boris N. Martinez. A power struggle was taking place around the time the disc message was sent.

## Peru
Juan Francisco Velasco Alvarado (1910 – 1977), was President from 1968 to 1975. General Velasco took power in a 1968 military coup. The coup named the new government the "Revolutionary Government of the Armed Forces." Velasco's rule, known as Velascato, was socialistic. He nationalized entire industries, and consolidated them into single government-run entities and increased government control by preventing any private activity in industry sectors. His government resorted to jailing, deporting and harassing suspected political opponents. He had close ties with The Soviet Union and Cuba. In 1975 he was deposed by a military coup.

## Phillipines
Ferdinand Emmanuel Edralín Marcos (1917 – 1989), was President of the Philippines from 1965 to 1986. Marcos became one of the most corrupt and fraudulent presidents ever seen. He took billions of dollars in foreign aid, kickbacks and patronage that allowed Marcos to rule until 1986. In 1972, he declared martial law. Rule was by executive decree, laws passed by a rubber stamp parliament. In 1986, an army coup, which turned into an uprising under Corazon Aquino, forced him to leave the country. He died in exile in the U.S. in 1989.

## Poland
Jorzy Michalowski was an ambassador sent by Poland to give a message for the disc. Poland was part of the communist bloc countries.

## Portugal
Américo de Deus Rodrigues Tomás (1894 – 1987), was President from 1958 to 1974. Tomás took power through corrupt elections. Tomás was overthrown and sent to exile in Brazil until 1980.

## Romania
Nicolae Ceausescu (1918 – 1989), was President from 1965 to 1989. With the death of Gheorghiu-Dej, Ceausescu succeeded to the leadership of Romania's Communist Party as first secretary. He challenged the dominance of the Soviet Union over Romania. In the 1960s, Ceausescu ended Romania's active participation in the Warsaw Pact military alliance and condemned the invasion of Czechoslovakia and the invasion of Afghanistan. His secret police maintained rigid controls over freedoms and they tolerated no internal dissent or opposition. He and his wife were eventually tried and convicted by a special military tribunal on charges of mass murder. They were then shot.

## Senegal
Léopold Sédar Senghor (1906 – 2001), was the first President of Senegal from 1960 to 1980. During his presidency, he set Senegal on the path toward a multi-party democracy. Senghor is also regarded as one of the greatest French-language poets of the 20th century. Senegal is one of the few African nations never to have had a coup. He developed African socialism called négritude philosophy. In developing this, he was assisted by Ousmane Tanor Dieng. In 1980, he retired in favour of his prime minister, Abdou Diouf.

## Sierra Leone
Siaka Probyn Stevens (1905 – 1988), was the Prime Minister from 1968 to 1971 and later President of the Republic of Sierra Leone from 1971 to 1985. He served as Chairman of the Organization of African Unity and drafted the Mano River Union, a three country federation of Sierra Leone, Liberia, and Guinea. He survived several coup and assassination attempts.

## South Africa
Jacobus Johannes ('Jim') Fouché (1898 – 1980), was President of South Africa from 1968 to 1975. As a defense minister, he built South Africa's military as a strong supporting structure for apartheid.

### Swaziland
Sobuza II (1899 – 1982), was King of Swaziland from 1921 to 1982. Sobhuza was King during the country's independence from Great Britain in 1968. Swaziland was a constitutional monarchy until 1973, when he made himself an absolute ruler. He married 70 wives and had 210 children.

### Thailand
Bhomuibol Adulyadej (Rama IX) (b. 1927), was King of Thailand from 1946 to the present. He is the world's longest-serving current head of state and the longest-serving monarch in Thai history. He supported military regimes, but then transitioned Thailand to a democracy in the 1990's. He was born in Cambridge, Massachusetts. He is very popular and is considered one of the wealthiest people in the world. He has used his wealth to help the country with various projects.

### Togo
General Gnassingbé Eyadéma, formerly Étienne Eyadéma (1937 – 2005), was President of Togo from 1967 to 2005. He participated in two successful military coups. He escaped several assassination attempts. In one instance, he carried the bullet removed by the surgeon as an amulet. He was the only person to survive a plane crash. He was considered to be a ruthless dictator to opponents.

### Trinidad and Tobago
Dr. Eric Eustace Williams (1911 – 1981), was the first Prime Minister of Trinidad and Tobago from 1956 to 1981. He received his doctorate in 1938 from Oxford University and was a scholar. He is often called the "Father of the Nation," and was one of the most significant leaders in the history of Trinidad and Tobago. Dr. Eric Williams considered himself a teacher, historian and a philosopher. Before and during his tenure as prime minister, he wrote many articles and books on philosophy, education, and politics.

### Tunisia
Habib Ben Ali Bourguiba (1903 – 2000), was President from 1957 to 1987. In 1957 a republic was proclaimed, abolishing the monarchy and giving Bourguiba a long tenure as President. He became a supporter of the United States and made many positive reforms for women. In 1987, his Prime Minister impeached him for health reasons.

## Turkey

Cevdet Sunay (1899 – 1982), was President of Turkey from 1966 to 1973. Cevdet Sunay was elected the fifth President in 1966. His office survived terrorist activity, riots, and potential coups. He served the constitutional term of seven years and then became a permanent senator.

## Upper Volta

Lieutenant-Colonel Aboubakar Sangoulé Lamizana (1916 – 2005), was President of Upper Volta from 1966 to 1980 and also Prime Minister from 1974 to 1978. Upper Volta achieved independence from the French Union in 1960 with Maurice Yaméogo as President. Yaméogo resigned and Lamizana served as nominal head of a military government and later became a dictator. Colonel Saye Zerbo overthrew President Lamizana in a bloodless military coup in 1980.

## Uruguay

Jorge Pacheco Areco (1920 – 1998), was President of Uruguay from 1967 to 1972. Pacheco governed by ignoring the Uruguayan Congress, declaring a state of emergency and martial law, violating human rights, as well as torturing opposing parties and election fraud.

## Vatican

Pope Paul VI, born Giovanni Battista Enrico Antonio Maria Montini (1897 – 1978), reigned as Pope of the Catholic Church and Sovereign of Vatican City from 1963 to 1978. Succeeding John XXIII, who had convened the Second Vatican Council, he presided over the majority of its sessions and oversaw the implementation of its decrees. Pope Paul VI is known for his encyclical Humanæ Vitæ (On the Regulation of Birth). This reaffirmed the Catholic Church's traditional condemnation of artificial birth control. Pope Paul VI became the first pope to visit six continents and was the most travelled pope in history to that time. In 1970 he was the target of a failed assassination attempt in the Phillipines.

## Vietnam

Nguyen Van Thieu (1923 – 2001), was the President of South Vietnam from 1965 to 1975. He joined the Viet Minh, led by He Chí Minh to liberate Vietnam from French colonialism. He abandoned them after they moved into the Soviet sphere of influence and embraced Communism. In 1975, North Vietnam invaded South Vietnam and Thieu was exiled.

## *Yugoslavia*
Josip Broz Tito (1892 – 1980), was President of the Socialist Republic of Yugoslavia from 1953 to 1980. Tito is known for organizing the anti-fascist resistance movement known as the Yugoslav Partisans along with resisting Soviet influence, and founding and promoting the Non-Aligned Movement.

## *Zambia*
Kenneth David Kaunda (b. 1924), was President of Zambia from 1964 to 1991. Kaunda helped found the Lubwa branch of the African National Congress. Kaunda became Prime Minister of Northern Rhodesia and then he became the first President of the newly independent Zambia. He now devotes time doing charity work for the anti-HIV/AIDS campaign.

# Appendix 2

## TREATY ON PRINCIPLES GOVERNING THE ACTIVITIES OF STATES IN THE EXPLORATION AND USE OF OUTER SPACE, INCLUDING THE MOON AND OTHER CELESTIAL BODIES
## (1967)

ENTERED INTO FORCE: 10 October 1967

The States Parties to this Treaty,

Inspired by the great prospects opening up before mankind as a result of man's entry into outer space,

Recognizing the common interest of all mankind in the progress of the exploration and use of outer space for peaceful purposes,

Believing that the exploration and use of outer space should be carried on for the benefit of all peoples irrespective of the degree of their economic or scientific development,

Desiring to contribute to broad international co-operation in the scientific as well as the legal aspects of the exploration and use of outer space for peaceful purposes,

Believing that such co-operation will contribute to the development of mutual understanding and to the strengthening of friendly relations between States and peoples,

Recalling resolution 1962 (XVIII), entitled "Declaration of Legal Principles Governing the Activities of States in the Exploration and Use of Outer Space", which was adopted unanimously by the United Nations General Assembly on 13 December 1963,

Recalling resolution 1884 (XVIII), calling upon States to refrain from placing in orbit around the earth any objects carrying nuclear weapons or any other kinds of weapons of mass destruction or from installing such weapons on celestial bodies, which was adopted unanimously by the United Nations General Assembly on 17 October 1963,

Taking account of United Nations General Assembly resolution 110 (II) of 3 November 1947, which condemned propaganda designed or likely to provoke or encourage any threat to the peace, breach of the peace or act of aggression, and considering that the aforementioned resolution is applicable to outer space,

Convinced that a Treaty on Principles Governing the Activities of States in the Exploration and Use of Outer Space, including the Moon and Other Celestial Bodies, will further the Purposes and Principles of the Charter of the United Nations,

Have agreed on the following:

## Article I

The exploration and use of outer space, including the moon and other celestial bodies, shall be carried out for the benefit and in the interests of all countries, irrespective of their degree of economic or scientific development, and shall be the province of all mankind.

Outer space, including the moon and other celestial bodies, shall be free for exploration and use by all States without discrimination of any kind, on a basis of equality and in accordance with international law, and there shall be free access to all areas of celestial bodies.

There shall be freedom of scientific investigation in outer space, including the moon and other celestial bodies, and States shall facilitate and encourage international co-operation in such investigation.

## Article II

Outer space, including the moon and other celestial bodies, is not subject to national appropriation by claim of sovereignty, by means of use or occupation, or by any other means.

## Article III

States Parties to the Treaty shall carry on activities in the exploration and use of outer space, including the moon and other celestial bodies, in accordance with international law, including the Charter of the United Nations, in the interest of maintaining international peace and security and promoting international co-operation and understanding.

## Article IV

States Parties to the Treaty undertake not to place in orbit around the earth any objects carrying nuclear weapons or any other kinds of weapons of mass destruction, instal such weapons on celestial bodies, or station such weapons in outer space in any other manner.

The moon and other celestial bodies shall be used by all States Parties to the Treaty exclusively for peaceful purposes. The establishment of military bases, installations and fortifications, the testing of any type of weapons and the conduct of military manoeuvres on celestial bodies shall be forbidden. The use of military personnel for scientific research or for any other peaceful purposes shall not be prohibited. The use of any equipment or facility necessary for peaceful exploration of the moon and other celestial bodies shall also not be prohibited.

## Article V

States Parties to the Treaty shall regard astronauts as envoys of mankind in outer space and shall render to them all possible assistance in the event of accident, distress, or emergency landing on the territory of another State Party or on the high seas. When astronauts make such a landing, they shall be safely and promptly returned to the State of registry of their space vehicle.

In carrying on activities in outer space and on celestial bodies, the astronauts of one State Party shall render all possible assistance to the astronauts of other States Parties.

States Parties to the Treaty shall immediately inform the other States Parties to the Treaty or the Secretary-General of the United Nations of any phenomena they discover in outer space, including the moon and other celestial bodies, which could constitute a danger to the life or health of astronauts.

## Article VI

States Parties to the Treaty shall bear international responsibility for national activities in outer space, including the moon and other celestial bodies, whether such activities are carried on by governmental agencies or by non-governmental entities, and for assuring that national activities are carried out in conformity with the provisions set forth in the present Treaty. The activities of non-governmental entities in outer space, including the moon and other celestial bodies, shall require authorization and continuing supervision by the appropriate State Party to the Treaty. When activities are carried on in outer space, including the moon and other celestial bodies, by an international organization, responsibility for compliance with this Treaty shall be borne both by the international organization and by the States Parties to the Treaty participating in such organization.

## Article VII

Each State Party to the Treaty that launches or procures the launching of an object into outer space, including the moon and other celestial bodies, and each State Party from whose territory or facility an object is launched, is internationally liable for damage to another State Party to the Treaty or to its natural or juridical persons by such object or its component parts on the Earth, in air space or in outer space, including the moon and other celestial bodies.

## Article VIII

A State Party to the Treaty on whose registry an object launched into outer space is carried shall retain jurisdiction and control over such object, and over any personnel thereof, while in outer space or on a celestial body. Ownership of objects launched into outer space, including objects landed or constructed on a celestial body, and of their component parts, is not affected by their presence in outer space or on a celestial body or by their return to the Earth. Such objects or component parts found beyond the limits of the State Party to the Treaty on whose registry they are carried shall be returned to that State Party, which shall, upon request, furnish identifying data prior to their return.

## Article IX

In the exploration and use of outer space, including the moon and other celestial bodies, States Parties to the Treaty shall be guided by the principle of co-operation and mutual assistance and shall conduct all their activities in outer space, including the moon and other celestial bodies, with due regard to the corresponding interests of all other States Parties to the Treaty. States Parties to the Treaty shall pursue studies of outer space, including the moon and other celestial bodies, and conduct exploration of them so as to avoid their harmful contamination and also adverse changes in the environment of the Earth resulting from the introduction of extraterrestrial matter and, where necessary, shall adopt appropriate measures for this purpose. If a State Party to the Treaty has reason to believe that an activity or experiment planned by it or its nationals in outer space, including the moon and other celestial bodies, would cause potentially harmful interference with activities of other States Parties in the peaceful exploration and use of outer space, including the moon and other celestial bodies, it shall undertake appropriate international consultations before proceeding with any such activity or experiment. A State Party to the Treaty which has reason to believe that an activity or experiment planned by another State Party in outer space, including the moon and other celestial bodies, would cause potentially harmful interference with activities in the peaceful exploration and use of outer space, including the moon and other celestial bodies, may request consultation concerning the activity or experiment.

## Article X

In order to promote international co-operation in the exploration and use of outer space, including the moon and other celestial bodies, in conformity with the purposes of this Treaty, the States Parties to the Treaty shall consider on a basis of equality any requests by other States Parties to the Treaty to be afforded an opportunity to observe the flight of space objects launched by those States.

The nature of such an opportunity for observation and the conditions under which it could be afforded shall be determined by agreement between the States concerned.

### Article XI

In order to promote international co-operation in the peaceful exploration and use of outer space, States Parties to the Treaty conducting activities in outer space, including the moon and other celestial bodies, agree to inform the Secretary-General of the United Nations as well as the public and the international scientific community, to the greatest extent feasible and practicable, of the nature, conduct, locations and results of such activities. On receiving the said information, the Secretary-General of the United Nations should be prepared to disseminate it immediately and effectively.

### Article XII

All stations, installations, equipment and space vehicles on the moon and other celestial bodies shall be open to representatives of other States Parties to the Treaty on a basis of reciprocity. Such representatives shall give reasonable advance notice of a projected visit, in order that appropriate consultations may be held and that maximum precautions may be taken to assure safety and to avoid interference with normal operations in the facility to be visited.

### Article XIII

The provisions of this Treaty shall apply to the activities of States Parties to the Treaty in the exploration and use of outer space, including the moon and other celestial bodies, whether such activities are carried on by a single State Party to the Treaty or jointly with other States, including cases where they are carried on within the framework of international inter-governmental organizations.

Any practical questions arising in connexion with activities carried on by international inter-governmental organizations in the exploration and use of outer space, including the moon and other celestial bodies, shall be resolved by the States Parties to the Treaty either with the appropriate international organization or with one or more States members of that international organization, which are Parties to this Treaty.

### Article XIV

1. This Treaty shall be open to all States for signature. Any State which does not sign this Treaty before its entry into force in accordance with paragraph 3 of this Article may accede to it at any time.

2. This Treaty shall be subject to ratification by signatory States. Instruments of ratification and instruments of accession shall be deposited with the Governments of the United Kingdom of Great Britain and Northern Ireland, the Union of Soviet Socialist Republics and the United States of America, which are hereby designated the Depositary Governments.

3. This Treaty shall enter into force upon the deposit of instruments of ratification by five Governments including the Governments designated as Depositary Governments under this Treaty.

4. For States whose instruments of ratification or accession are deposited subsequent to the entry into force of this Treaty, it shall enter into force on the date of the deposit of their instruments of ratification or accession.

5. The Depositary Governments shall promptly inform all signatory and acceding States of the date of each signature, the date of deposit of each instrument of ratification of and accession to this Treaty, the date of its entry into force and other notices.

6. This Treaty shall be registered by the Depositary Governments pursuant to Article 102 of the Charter of the United Nations.

## Article XV

Any State Party to the Treaty may propose amendments to this Treaty. Amendments shall enter into force for each State Party to the Treaty accepting the amendments upon their acceptance by a majority of the States Parties to the Treaty and thereafter for each remaining State Party to the Treaty on the date of acceptance by it.

## Article XVI

Any State Party to the Treaty may give notice of its withdrawal from the Treaty one year after its entry into force by written notification to the Depositary Governments. Such withdrawal shall take effect one year from the date of receipt of this notification.

## Article XVII

This Treaty, of which the English, Russian, French, Spanish and Chinese texts are equally authentic, shall be deposited in the archives of the Depositary Governments. Duly certified copies of this Treaty shall be transmitted by the Depositary Governments to the Governments of the signatory and acceding States.

In witness whereof the undersigned, duly authorised, have signed this Treaty.

Done in triplicate, at the cities of London, Moscow and Washington, the twenty-seventh day of January, one thousand nine hundred and sixty-seven.

# Appendix 3

U.S. Patent for the Silicon Disc by John L.Sprague, Robert S. Pepper, Eugene P. Donovan, and Frederick W. Howe invented the disc, with the U.S. patent number 3,607,347.

## BACKGROUND OF THE INVENTION

This invention pertains to data reduction and storage and more particularly to high density storage of alphanumerical data in a substantially permanent medium.

In the prior art, reduction and storage of graphical or alphanumerical data are generally provided by microfilm techniques which although satisfactory for many uses, is limited in regards to achievable information density, and is susceptible to deterioration due to environmental conditions such that it fails to provide archival permanence.

It is an object of this invention to provide extremely high density storage of alphanumerical information.

It is another object of this invention to provide a recording medium suitable for permanent high density storage of alphanumerical data.

It is a further object of this invention to provide a high density storage device wherein alphanumerical information is recorded in microscopic size by line openings extended through a thin overlayer to a contrasting underlayer.

It is a still further object of this invention to provide a method of reducing and storing information within a permanent medium.

It is a further object of this invention to provide a method of storing information in microscopic size within a permanent medium by photoetching fine-line geometries through a thin overlying coating to an underlying layer.

## SUMMARY OF THE INVENTION

In accordance with the invention the storage device, which provides substantially permanent storage of data in microscopic size, comprises a substrate having at least one major surface, a thin overlayer of substantially permanent material disposed on said surface, and fine-line openings disposed in said overlayer and arranged to provide a representation of graphical information in microscopic size.

In a more limited sense, the storage device comprises a substantially planar substrate having a thin film layer disposed on a substantially smooth surface thereof, said thin film providing contrast to said substrate, a plurality of fine-line openings disposed in said surface coating so as to expose said substrate surface, and said openings arranged in the form of alphanumerical indicia of microscopic size.

Broadly, in the method of the invention, high density data storage is provided by removing portions of a thin film surface layer to expose an underlying substrate in a representative image in microscopic size of alphanumerical information so as to provide a microscopic representation of said information

In a preferred embodiment, the method of storing the information comprises forming a substrate having at least one smooth major surface, forming a thin film of durable material on said surface, depositing a radiation sensitive coating on said thin film, forming a reduced image of alphanumerical information on said coating by exposure to actinic radiation, removing reacted portions of said coating to leave fine-line openings therein which expose said thin film in a representative image in conformance with the line indicia of said information, etching away the exposed portions of said thin film to provide fine-line openings extending to said substrate, and removing said coating to provide a substantially permanent record in microscopic size of said indicia with said thin film.

BRIEF DESCRIPTION OF THE DRAWING

FIG. 1 is a plan view of a storage plate provided in accordance with the invention;

FIG. 2 is a plan view magnified many times of a small information area of the plate shown in FIG. 1; and

FIG. 3 is a view in section of the record taken along the lines 3--3 of FIG. 2.

DESCRIPTION OF THE PREFERRED EMBODIMENTS

In FIG. 1, a plate 10 is shown having a large amount of information or data 12 permanently recorded in alphanumerical form and microscopic size on its upper surface 14. This is shown more clearly in FIG. 2 wherein a small portion 16 of plate 10 is magnified many times to show the indicia 18 in approximately it original size.

Preferably, plate 10 is a substantially flat sheet having a body or substrate 20 of durable material for example, an inorganic material such as silicon or tantalum or the like and having a thin film surface layer 22 of highly durable material which provides a contrast and has high adherence to the substrate surface 24. For example silicon dioxide, or other compounds of the substrate which are grown or formed in situ on the substrate provide suitable adherence.

The information, or that is, the microscopic sized, representative indicia or characters 18 are formed by fine-line openings or trenchlike cuts 26 which extend through thin film 22 to surface 24 so as to provide a representative image of the information permanently stored in plate 10.

In the preferred method of the invention, a substrate 20 of inorganic material such as silicon is first prepared by conventional means with a highly polished upper surface 24. Thereafter a thin film surface coating 22 of silicon dioxide or the like is grown on the substrate surface. A photosensitive coating (not shown) is then deposited on the thin film layer. The alphanumerical information is photographically reduced, and the resulting positive image exposed on the photosensitive coating by ultraviolet light. Unexposed portions of the coating are then removed using conventional photoetching techniques to leave fine-line openings in the photoresist. Exposed portions of the thin film are then etched away by an acid solution, and finally, the photosensitive coating is thereafter removed to provide storage record plate 10 having data recorded by the fine-line openings of the thin film which are arranged to represent in microscopic size, the alphanumerical characters of the original documents.

Preferably, surface 24 is a smooth polished surface having selected radiation properties so as to provide contrast to the thin film. Hence, the underlying surface 24 and the thin film surface 14 should be sufficiently different in reflectivity, transmission or absorption of radiation, or in combinations of these, as to clearly reveal the characters when exposed to selected radiation, For optical retrieval of the information, optical contrast between the substrate and thin film is preferred. In addition to the cited characteristics, where the thin film is transparent, as in the case of silicon dioxide, the reflectivity of the underlying substrate must be low, or that is, in the range of the reflectivity of the thin film surface so that visible interference colors can result.

In the case of transparent films, the color results from the elimination or subtraction of particular frequencies by interference. Thus, with white light illumination of the unit, light having a wavelength $\lambda$ will be blanked out, or subtracted from, the reflected radiation depending upon film thickness. For the silicon-silicon dioxide system 900-1,000 Angstroms gives a characteristic blue-purple color when exposed to white light, while other colors of the spectrum results with thicker coatings. Generally 1,000--5,000 Angstroms coating thicknesses will be suitable, however, coatings in the 1,000 Angstrom range are preferred for the fine-line geometries necessary for maximum information density.

Monochromatic light can also be utilized for illumination and retrieval of information from this record system, since for example, if the slice is viewed with monochromatic light whose wavelength is in the interference or blanking range of the transparent coating, the coating will appear black while the exposed surface portions of the substrate (the characters) will reflect the source illumination.

Many different thin film and substrate materials can be employed. Preferably, both would be of inorganic material having high durability so as to provide archival permanence, that is, the structure should have good mechanical, chemical and thermal stability. Preferably, the thin film material should exhibit an exceptionally firm bond, to the substrate, as for example, the cohesive bond to be expected from thin coatings grown by chemical reaction with the substrate. However, films formed or deposited by other means such as vacuum deposition and the like may also be suitable.

Information may be recorded on one or more of the surfaces of the plate; for example, on both its planar surfaces. The data may be representative of nonrelated documents which are composed in side by side relation on the plate surface, or could provide a continuous narrative or account etc.

In a specific example, 74 printed sheets of approximately 81/2" .times.11" size were reduced approximately 200 times in size and stored in a 0.21 square area of a single 1- 1/2-inch diameter, 14 mil thick oxidized slice of silicon. Each printed sheet was first copied by electrostatic means, and a 1 to 1 positive transparency then made by means of the Diazo process. Groups of approximately 6 to 9 of the transparencies were then mounted in a side-by-side relationship and photographically reduced by 20 X and recorded on high resolution photographic plates by means of a photographic system.

The initial composite plates were then further reduced by about 10 times and composed in a side-by-side relation on a single high resolution photographic plate to provide a positive image mask. This was accomplished by photographically exposing each initial composite negative at selected coordinates of the final photographic plate by means of a step and repeat camera; that is, a different composite plate was exposed at each step of the camera. The final composed image was then transferred to the oxidized silicon slice by photolithographic techniques.

A 1,000 Angstrom thick surface film of silicon dioxide was first thermally grown on the slice by conventional semiconductor techniques of heating the slice to approximately 1,100.degree. C in an oxygen atmosphere. The slice was then coated with very thin KTFR photoresist, for example, a 5,000 Angstrom thick coating. The final composite was then disposed over the photoresist and the combination exposed to ultra violet light in the conventional manner. Thereafter, the photoresist was washed away in the developed or that is, nonpolymerized areas so as to provide fine-line openings in the resist coating in conformance with the printed characters of the original documents.

The unit was then treated with an ammonium fluoride buffered solution of hydrofluoric acid to etch through the silicon oxide exposed in the openings of the photoresist coating. The solution is chosen to rapidly etch the thin film, and to not attack or only very slowly etch the substrate. Finally, the unit was completed by stripping the photoresist with sulfuric acid or the like. This provided a silicon slice having approximately 74 images of the original documents within a 0.21-inch square area of the slice.

The final composite plate in this case was a positive image; that is, the dark lines of the composite presented, in highly reduced size, the printed lines of the original information. Hence, in this case, the openings in the photoresist corresponded to the printed lines of the original. However, a negative image could also be utilized to result in raised lettering that is, in relief. Moreover, other types of photoresist could be employed with a positive image or the like to permit removal of the coating around the lettering etc.

Other means of providing the initial composite plate can be utilized. For example, reversal film can be employed to provide a direct positive plate from the original document prior to the initial reduction and composition step. Moreover, either or both the initial and final composite plates may be made by reduction and composing directly from the original document. Additionally, different photographic reduction techniques may also be employed to provide the final photographic plate of the specific example, and it should be understood that the step and compose technique may be employed to provide a composite directly on the photoresist so as to eliminate contact printing of the photoresist.

As previously indicated, many different substrates and surface films may be utilized to store graphic line information. The underlying substrate should provide support and contrast to the thin film. Additional support and permanence can also be achieved by mounting the substrate on a supporting base.

Many different contrasting properties can be employed. That is, transparent thin films can be employed on an opaque substrate as described in preferred embodiment. Opaque films of different color than the substrate can be utilized. Additionally, a transparent substrate may be suitable, for example, aluminum, gold or other thin metallic films on plastics such as Mylar, or glass or quartz would be useful. Chromium, deposited by sputtering or the like, on glass or quartz will provide a highly permanent medium.

For high speed retrieval of the informations, such as for use in computers or the like, flexible substrates would be desirable. For example, metal thin films on plastics or contrasting metal layers would be suitable.

Of course, any graphic representation may be reduced and stored in the indicated manner. For example, three dimensional images may be provided by providing two superimposed images etched to different depths. Advantageously, three-dimensional displays may also be stored in two side-by-side patterns which are later composed by the retrieval system.

Information retrieval can be accomplished in many different ways. For example, for retrieval from a Si-SiO.sub.2 record, optical means utilizing reflected radiation is suitable. A microscope or an optical comparator can be utilized to a view a magnified image of the stored data. In this case, the instrument can be directed to selected coordinates of the record by manual or electronic means or the like.

Direct viewing or projection-type arrangements may be employed for records designed for viewing by either reflected or transmitted illumination. The latter would require records having a substantially opaque thin film and a substantially transparent substrate, such as for example, aluminum or other metallic films on glass or quartz or the like.

Other forms of retrieval, such as electron scanning can be utilized. For example, a flying spot scanner and electron microscope techniques can be employed for electronic read out and the like. That is, both primary and secondary interaction with the stored medium may be employed. For electron scanning or the like, the contrast between the thin film and substrate must be applicable to that radiation rather than optical radiation as previously indicated.

Thin films of inorganic materials having a high melting point, for example of metals such as chromium or refractory materials such as oxide or the like are preferable. Of course, the substrate can be of any material which will not react with the film and which provides mechanical support and contrast to it. Moreover, the substrate itself may be laminated to or mounted on other materials such as metals or the like to protect the unit from mechanical stress, etc.

There is a fundamental limitation due to diffraction scattering of light such that images down to about one-half micron can only be formed with ultraviolet light. For finer-line geometries and greater reduction factors, electron beam techniques can be employed. For example, the image can be written in highly reduced form on suitable photoresist by a controlled electron beam, or the beam can be used directly on the thin film to remove or cut the film, for example by evaporation of a metal film.

Hence many different embodiments may be realized without departing from the spirit and scope of the invention and it is to be understood that the invention is not to be limited except as in the appended claims.

# Appendix 4

# Apollo Missions

### Apollo 7
Saturn 1B
October 11-22, 1968
Walter M. Schirra Jr. (commander), Donn F. Eisele (CM pilot), R. Walter Cunningham (LM pilot)

### Apollo 8
Saturn V (AS-503, CSM-103)
December 21-27, 1968
Frank Borman (commander), James A. Lovell Jr. (CM pilot), William A. Anders (LM pilot)
*Made 10 orbits around the moon*

### Apollo 9 (Gumdrop and Spider)
Saturn V (AS-504, SM-104, CM-104, LM-3)
March 03-13, 1969
James A. McDivitt (commander), David R. Scott (CM pilot), Russell L. Schweickart (LM pilot)
*Lunar Module test*

### Apollo 10 (Charlie Brown and Snoopy)
Saturn V (AS-505, SM-106, CM-106, LM-4)
May 18-26, 1969
Thomas P. Stafford (commander), John W. Young (CM pilot), Eugene A. Cernan (LM pilot)
*Lunar Module test near Moon*

### Apollo 11 (Columbia and Eagle)
Saturn V (AS-506, SM-107, CM-107, LM-5)
July 16-24, 1969
Neil A. Armstrong* (commander), Michael Collins (CM pilot), Edwin E. (Buzz) Aldrin Jr.* (LM pilot)
*First lunar landing*

### Apollo 12 (Yankee Clipper and Intrepid)
Saturn V
November 14-24, 1969
Charles Conrad Jr.* (commander), Richard F. Gordon Jr. (CM pilot), Alan L. Bean* (LM pilot)
*Second lunar landing*

### Apollo 13 (Odyssey and Aquarius)
Saturn V
April 11-17, 1970
James A. Lovell Jr. (commander), John L. Swigert Jr. (CM pilot), Fred W. Haise Jr. (LM pilot)
*Mission aborted due to emergency.*

### Apollo 14 (Kitty Hawk and Antares)
Saturn V (AS-509, SM-110, CM-110, LM-8)
January 31-February 09, 1971
Alan B. Shepard Jr.* (commander), Stuart A. Roosa (CM pilot), Edgar D. Mitchell* (LM pilot)
*Third lunar landing*

### Apollo 15 (Endeavor and Falcon)
Saturn V
July 26-August 07, 1971
David R. Scott* (commander), Alfred M. Worden (CM pilot), James B. Irwin* (LM pilot)
*Fourth lunar landing*

### Apollo 16 (Casper and Orion)
Saturn V
April 16-27, 1972
John W. Young* (commander), Thomas K. Mattingly II (CM pilot), Charles M. Duke Jr.* (LM pilot)
*Fifth lunar landing*

### Apollo 17 (America and Challenger)
Saturn V
December 07-19, 1972
Eugene A. Cernan* (commander), Ronald E. Evans (CM pilot), Harrison H. Schmitt *(LM pilot)
*Sixth lunar landing*

*indicates one of 12 Moonwalker astronauts

# Appendix 5

### Notes:
1. "It's pretty much without color. It's gray, and it's very white, chalky gray, as you look into the zero-phase line. And it's considerably darker gray, more like ashen gray, as you look out ninety degrees to the Sun." (Neil Armstrong, First Man, p.485).
2. "The raising of an American flag would seem most undesirable from this standpoint, since such an action has historically symbolized conquest and territorial acquisition." (U. Alexis Johnson, General records, Department of State, Central Foreign Policy Files, 1967-1969, Political and Defense, Box 2521).
3. "My job was to get the flag there. I was less concerned about whether that was the right artifact to place. I let other, wiser minds than mine make those kinds of decisions, and I had no problem with it. (Neil Armstrong, First Man, p.395)."
4. "In view of our total ignorance of this project… and King's apparent keen interest, would appreciate any information you can provide concerning NASA invitation to send message . . . number of countries responding . . . methods of recording and method of deposit on the Moon." (Telegram, The Papers of Thomas O. Paine, Manuscript Division, Library of Congress, Box 45).
5. "We who first walk the surface of the Moon leave this plaque to commemorate our journey and to mark Man's progress in his continuing quest for a more complete understanding of the universe. We came as envoys of mankind, exploring the moon for the benefit of all peoples. May this voyage not only illuminate the mysteries of the universe, but unite us in the search for truth and understanding on our own planet." (U. Alexis Johnson, General records, Department of State, Central Foreign Policy Files, 1967-1969, Political and Defense, Box 2521).
6. "Today the whole world knows the secret you have helped us keep for many months. I am pleased to be able to add that the warlords of Japan now know its effects better than we ourselves. The atomic bomb, which you have helped to develop with high devotion to patriotic duty, is the most devastating military weapon that any country has ever been able to turn against its enemy. No one of you has worked on the entire project or known the whole story. Each of you has done his own job and kept his own secret, so today I speak for a grateful nation when I say congratulations, and thank you all. I hope you will continue to keep the secrets you have kept so well. The need for security and for continued effort is as fully great now as it ever was. We are proud of every one of you." (Robert P. Patterson, Sprague Electric Company News Release, July 14, 1969)
7. "Crash program is an understatement. We had almost no time to put this together!" John L. Sprague, author interview.

8. "It was a rush to get it done. We slept on lab benches for two days in row." (Ray Carswell, Spaceport News article, Apollo 11 Goodwill Messages Remembered, unknown date).
9. "The last minute requirement of a complete redo to add more world messages came as an unexpected shock."(John Sprague, author interview)
10. "We were unencumbered by all the bureaucratic layers of a larger organization. Decisions could be made and implemented instantaneously, and no one even considered failure." (John Sprague, author interview)
11. "Ed (White) was a good friend of mine."(Buzz Aldrin, author interview)
12. "And I think everybody shares that observation, and I don't know why you have that impression, but it's so small, it's very colorful—you know, you see an ocean and gaseous layer, a little bit, just a tiny bit, of atmosphere around it, and compared with all the other celestial objects, which in many cases are much more massive, more terrifying, it just looks like it couldn't put up a very good defense against a celestial onslaught."(Neil Armstrong, NASA Johnson Space Center, Oral history project transcript, 2001).
13. "Trying to get into a pretty tight spot probably wouldn't be fun. Also, the area was coming up quickly, and it soon became obvious that I could not stop short enough to find a safe landing spot…" (Neil Armstrong, First Man, p.466)
14. "Fortunately, there were no really harrowing parts of the flight. The most difficult part, from my perspective, and the one that gave me the most pause, was the final descent to landing. That was far and away the most complex part of the flight. The systems were very heavily loaded at that time. The unknowns were rampant. The systems in this mode had only been tested on Earth and never in the real environment. There were just a thousand things to worry about in the final descent. It was hardest for the system and it was hardest for the crews to complete that part of the flight successfully."(Neil Armstrong, NASA Johnson Space Center, Oral history project transcript, 2001)
15. "I was absolutely dumbfounded when I shut the rocket engine off and the particles that were going out radially from the bottom of the engine fell all the way out over the horizon, and when I shut the engine off, they just raced out over the horizon and instantaneously disappeared, you know, just like it had been shut off for a week. That was remarkable. I'd never seen that. I'd never seen anything like that. And logic says, yes, that's the way it ought to be there, but I hadn't thought about it and I was surprised." (Neil Armstrong, NASA Johnson Space Center, Oral history project transcript, 2001)
16. Aldrin Communion quotes: Author interview and flown card (see ill).
17. "In the one-sixth gravity of the Moon, the wine curled slowly and gracefully up the side of the cup. It was interesting to think that the very first liquid ever poured on the Moon, and the first food eaten there, were Communion elements." Buzz Aldrin, Men from Earth, p.240 and Time magazine, Nov. 15, 1971.
18. "That's one small step for (a) man, one giant leap for mankind", First Man, p.493.

19. "I was surprised by the apparent closeness of the horizon. I was surprised by the trajectory of dust that you kicked up with your boot, and I was surprised that even though logic would have told me that there shouldn't be any, there was no dust when you kicked. You never had a cloud of dust there. That's a product of having an atmosphere, and when you don't have an atmosphere, you don't have any clouds of dust." (Neil Armstrong, NASA Johnson Space Center, Oral history project transcript, 2001).
20. "Hello, Neil and Buzz. I'm talking to you by telephone from the Oval Room at the White House and this certainly has to be the most historic phone call ever made. For one priceless moment in the whole history of man, all the people on this Earth are truly one." (Richard Nixon, First Man, p.505).
21. "We had forgotten about this up to this point. And I don't think we really wanted to totally openly talk about what it was. So it was sort of guarded. And I knew what he (Armstrong) was talking about." (Buzz Aldrin, (Jones, Eric P., Apollo Lunar Surface Journal, Apollo 11 Technical Crew debriefing, July 31, 1969).
22. "We were so busy that I was halfway up the ladder before Neil asked me if I had remembered to leave the mementos we had brought along. I had completely forgotten. What we had hoped to make into a brief ceremony, had there been time, ended almost as an afterthought. I reached into my shoulder pocket, pulled the packet out and tossed it onto the surface." Buzz Aldrin, (Return to Earth p. 238 and author interview).
23. "Human character this is the area where we've made the least progress learning more about the brain, about our behavior and the ways we relate to one another. I think that's the most important direction we can take in the next twenty years; basically to begin to understand ourselves." (Neil Armstrong, quoted by Sydney J. Harris, in an article entitled: "Character: our next Frontier," Chicago Daily News, newspaper article ca. 1971).
24. "If you believe the Earth's increasing appetite for energy and the suspected future decrease in available energy will create an ever more severe problem for our Earth's future, you will find this proposal worthy of careful examination." (Neil Armstrong, Foreword, Return to the Moon, Schmitt).
25. "The rocket worked perfectly except for landing on the wrong planet." (Wernher von Braun, September 1944, (University of Michigan, the Past and future of rocket engine propulsion).
26. "For, in the final analysis, our most basic common link is that we all inhabit this small planet. We all breathe the same air. We all cherish our children's future. And we are all mortal." (John F. Kennedy, Commencement Address at American University, Washington, D.C., June 10, 1963)
27. "We will see a manned scientific base being built on the Moon. It'll be a scientific station manned by an international crew, very much like the Antarctic station. But there is a much more important question than whether man will be able to live on the Moon. We have to ask ourselves whether man will be able to live together down here on Earth." (Neil Armstrong, First Man p. 581).
28. Quotations from the silicon disc goodwill messages are referenced on the page that the quote appears, (Sprague Electric and NASA press release 69-83F, July 13, 1969).

# Acknowledgements

I have been a space enthusiast and historian for over 12 years. First and foremost, I would like to thank author and journalist Andrew Chaikin. His work on *A Man on the Moon* inspired me to learn more about the Apollo program. He is a gifted writer and was of immense help in making this project possible.

Buzz Aldrin is a great person to meet, along with his stepdaughter, Lisa Cannon. Buzz was generous enough to allow me access to his thoughts and stories about the Moon landing and his recollection of the disc itself. His eyes showed me that he has deep respect for the messages he left on the Moon. When he saw the title of this book, he stated, "This really is the untold story!" He motivated me to do things as accurately as possible for posterity.

James R. Hansen was also a great help. *First Man* is an extraordinary wealth of information. Hansen also helped me with key points that he confirmed with Neil Armstrong. Few men are as humble and approachable as Hansen. Two important historians are Eric Jones and Robert Pearlman (owner of collectSPACE.com). Their special insights into the Apollo program were instrumental to many parts of this book. *The Lunar Surface Journal* is an extraordinary resource that Eric Jones kindly allowed me to use here.

John Sprague gave me the entire story of the manufacturing process of the silicon disc. His firsthand account of actually etching the messages on the disc is truly a remarkable story. His information made me confident that this project could be successful. He is a kind man and belongs to a generation of engineers that I admire greatly. We can learn from their stories to strive toward missions to Mars and beyond.

Shaun Nicholson (owner of Tonerhaus.com) was also a great resource. He is a former TRW engineer who also worked on NASA radar projects in his career. His superior writing skills and clever ideas made for a more energetic reading style of the book.

Paul R. McHugh, former director of the Johns Hopkins psychiatry department, trained me as a resident psychiatrist. Being a psychiatrist can bring a new dimension of thinking in the context of any project. More psychiatric knowledge will be needed for our species to successfully venture into space and endure long periods of isolation.

Angela Farley is a gifted cover artist in Kansas City. It was her idea to make a grand presentation of the disc rising majestically from behind the Earth. Michele Rook is a lively, hard-working graphic designer. Her layout of the messages from world leaders is quite impressive and gives a dramatic and equal representation in the true spirit of the disc's intention. The women at Leathers Publishing have taken a giant leap forward in redesigning their operation. Barbra Thomson and Madlyn Davis are thoughtful and encouraging people.

Ken Havekotte was a great source of information and his collection is quite extraordinary. A silicon disc is now on display at the Kansas Cosmosphere and Space Center (KCSC) in Hutchinson, Kansas. I would like to thank Jim Remar and Meredith Miller for their kind help at the KCSC. Muriel Adams is an influential person in my life.

# Acknowledgements

Mathew Hogan of Maryland is an incredible researcher. His dry sense of humor and energetic searches kept me going. He e-mailed me with "bingo" when he found the archive containing information about the silicon disc and all the goodwill messages at the Library of Congress. That was a triumphant day. Susan Strange then photographed the archive for the book.

Sheila Sonnenschein and Walt Klein helped with editing and contributed to some big changes in the book's structure. Mike Constantine of Moonpans.com gave me the extraordinary full body photo of Neil Armstrong . He sells some truly magnificent posters. I very briefly met Armstong in September, 2007. We discussed the ejection seat of the LLTV. Joseph DeCuyper, Jr. provided some important advice and media contacts. R and R Enterprises (rrauction.com) provided ad space. They sell high quality astronaut signed material.

James Rendina and John Bingham stayed up all night to photograph the messages on the disc. That posed a more difficult job than I expected. These veteran microscope guys were patient and fun to work with. When we first saw the messages, we were astonished. I figured other people would want to see them, too.

My parents, Farhat and Hafiz Rahman immigrated to the U.S. from Pakistan. My father's scientific mind inspired me to learn everything I could. My mother brought out the best in my abilities. They took me on trips to the Kennedy Space Center and Kansas Cosmosphere. They inspired me to learn more about the universe. My brother, Jehan and my stepmother, Sheerin and I once painted a model of a Saturn V rocket when I was a child. Jehan and I pretended to be astronauts and wore flight suits from the NASA gift shop. Sheerin kept me from failing math in sixth grade. Gregory Santoscoy is an encyclopedia of wisdom.

My immediate family tolerated a lot of lengthy time spent by me on the computer and traveling to make this book possible. My wife, Stephanie, is a pragmatic and loving woman with a great soul. Our beautiful and energetic children, Jacqueline and Alec, keep things lively.

I hope this book fosters ideas of tolerance, peace and hope for the future. The messages on the silicon disc warn of our own self-destructive tendencies as human beings. I see this every day as a psychiatrist. We all see it in regional conflicts on the news. We must keep in mind that all of the world saw something good for humanity with the Apollo 11 mission.

That summer day in 1969 changed human history and made us unite. As man continues to place pressure on this finite planet, he will run out of resources. People will then reach for the stars to find answers to the most perplexing problems. This will require careful planning and cooperation. Hopefully mankind will find that spark of unification again — perhaps with another interplanetary voyage or another scientific adventure.

# *Bibliography*

## *Documents*

Jones, Eric P., *Apollo Lunar Surface Journal*

NASA Johnson Space Center, Oral history project transcript of Neil Armstrong, by Dr. Stephen E. Ambrose and Dr. Douglas Brinkley, September 19, 2001, Houston, TX.

NASA Press release 69-83F, July 13, 1969

NASA Fact Sheet, MR-5, Apollo 11

Sprague Electric Company News Release, July 14, 1969

The papers of Samuel C. Phillips, Manuscript Division, Library of Congress, Box 110,121

General Records of the Department of State, Central Foreign Policy Files, 1967-1969, Science, Box 3013

General Records of the Department of State, Central Foreign Policy Files, 1967-1969, Political and Defense, Box 2521

The Papers of Thomas O. Paine, Manuscript Division, Library of Congress, Box 45

Platoff, Anne, M., Hernandez Engineering, *Where No Flag has Gone Before*, North American Vexillogical Association, Lyndon B. Johnson Space Center, NASA contract no. NAS9-18263, October 1992.

Silicon Disc, Sprague Electric Company, July 1969

Sheer, Julian, Obituary, written by Andrew Chaikin

Spaceport News, *Apollo 11 Goodwill Messages Remembered*

AAPG Field Trip Guidebook, *Wildcatting on the Moon*. NASA Space Center, Houston, TX, by Harrison Scmitt.

## Newspapers and Periodicals

Houston Chronicle, October 3, 2006, "One Small Step for Clarity."

New York Times, July 15, 1969, page 20.

The Washington Post, Times Herald (1959-1973); Jul 15, 1969

Excerpts from Sprague Electric LOG, Vol. VIII, # 2, 8/25/1945

## Books

Aldrin, Buzz, and Malcolm McConnell. *Men from Earth*. 2nd ed. New York: Bantam and Falcon Books, 1991.

Aldrin, Edwin E. Jr, with Wayne Warga. *Return to Earth*. New York: Random House, 1973.

Chaikin, Andrew. *A Man on the Moon*. New York and London: Penguin Group, 1994.

Collins, Michael, *Carrying the Fire: An Astronaut's Journeys*. New York: Farrar, Strauss, and Giroux, 1974.

Hansen, James R., *First Man*. New York: Simon and Schuster, 2005.

Light, Michael, *Full Moon*, New York: Knopf, 1999

Murray, Charles and Catherine Bly, Cox, *Apollo, The Race to the Moon*. New York: Simon and Schuster, 1989.

Schmitt, Harrison, H., *Return to the Moon*. New York, Copernicus Books and Praxis Publishing, Ltd, 2006.

## E-mail and correspondence with author

Adams, Muriel, Overland Park, KS

Brammley, Paul, London, England

Brunner, Lisa, Kansas City, MO, Legal Counsel — Husch and Eppenberger

Bingham, John, Leawood, KS

Chaikin, Andrew, Arlington, MA

Constantine Mike, United Kingdom

Cornish, Scott, Houston, TX

DeCuyper, Jr., Joseph Y., Gladstone, MO

Hansen, James, R., Auburn, AL

Havekotte, Ken, Merritt Island, FL

James Rendina, Kansas City, MO

Jones, Eric, P., Australia

Lebendiger, Gary, A.H., Atlanta, GA

McCleod, Bob, Atlanta, GA

Miller, Meredith, Kansas Cosmosphere and Space Center, Hutchinson, KS

Nicholson, Shaun R., San Diego, CA

Pearlman, Robert, *collectSPACE.com*, Houston, TX

Pressley, Heather

Quigley, Tara, Overland Park, KS

Remar, Jim, Kansas Cosmosphere and Space Center, Hutchinson, KS

Santoscoy, Gregory, Wichita, KS

Sonnenschein, Sheila, Kansas City

Sprague, John, Williamstown, MA

## Book Design

Cover and Jacket Design — Angela Farley, Kansas City, MO

Interior Graphic Design — Michele Rook, Overland Park, KS

Microscopic Photography — James Rendina, Kansas City, MO

Microscopes — John Bingham, Leawood, KS

Moonpan Image of Neil Armstrong — Courtesy of *Moonpans.com*

Photographs — Courtesy of NASA

## Interviews

Author interview with Buzz Aldrin, August 17, 2007

Author interview with John Sprague, March, 2007

## Researchers

Hogan, Mathew, Maryland

Strange, Susan, Washington, D.C.

## Websites

www.buzzaldrin.com

www.collectSPACE.com

www.space.com

www.wikipedia.com